Fault Finding & Diagnosis

Malcolm Doughton and John Hooper

CENGAGE
Learning

Australia • Brazil • Japan • Korea • Mexico • Singapore • Spain • United Kingdom • United States

Fault Finding & Diagnosis
Malcolm Doughton and John Hooper

Publishing Director: Linden Harris

Commissioning Editor: Lucy Mills

Editorial Assistant: Claire Napoli

Project Editor: Alison Cooke

Production Controller: Eyvett Davis

Marketing Manager: Lauren Mottram

Typesetter: S4Carlisle Publishing Services

Cover design: HCT Creative

Text design: Design Deluxe

For product information and technology assistance,
contact **emea.info@cengage.com**.
For permission to use material from this text or product,
and for permission queries,
email **emea.permissions@cengage.com**.

British Library Cataloguing-in-Publication Data
A catalogue record for this book is available from the British Library.

ISBN: 978-1-4080-3996-0

Cengage Learning EMEA
Cheriton House, North Way, Andover, Hampshire, SP10 5BE, United Kingdom

Cengage Learning products are represented in Canada by Nelson Education Ltd.

For your lifelong learning solutions, visit **www.cengage.co.uk**

Purchase your next print book, e-book or e-chapter at **www.cengagebrain.com**

Printed in Malta by Melita Press
1 2 3 4 5 6 7 8 9 10 – 14 13 12

Dedication

This series of study books is dedicated to the memory of Ted Stocks whose original concept, and his publication of the first open learning material specifically for electrical installation courses, forms the basis for these publications. His contribution to training has been an inspiration and formed a solid base for many electricians practising their craft today.

The Electrical Installation Series

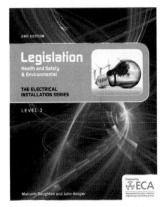

Legislation: Health and
Safety & Environmental

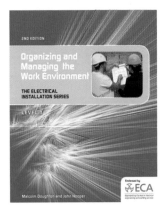

Organizing and Managing
the Work Environment

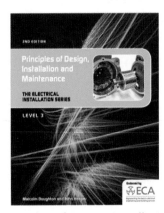

Principles of Design Installation
and Maintenance

Installing Wiring Systems

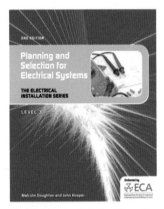

Planning and Selection for
Electrical Systems

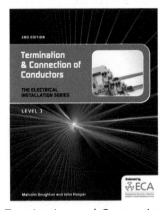

Termination and Connection
of Conductors

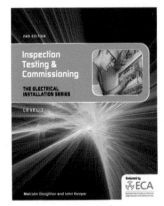

Inspection Testing
and Commissioning

Maintaining Electrotechnical
Systems

Contents

About the authors

Malcolm Doughton

Malcolm Doughton, I.Eng, MIET, LCG, has experience in all aspects of electrical contracting and has provided training to heavy current electrical engineering to HNC level. He currently provides training on all aspects of electrical installations, inspection, testing, and certification, health and safety, PAT and solar photovoltaic installations. In addition, Malcolm provides numerous technical articles and is currently managing director of an electrical consultancy and training company.

John Hooper

John Hooper spent many years teaching a diverse range of electrical and electronic subjects from craft level up to foundation degree level. Subjects taught include: Electrical Technology, Engineering Maths, Instrumentation, P.L.C.s, Digital, Power and Microelectronic Systems. John has also taught various electrical engineering subjects at both Toyota and JCB. Prior to lecturing in further and higher education he had a varied career in both electrical engineering and electrical installations.

Acknowledgements

The authors and publisher would like to thank Chris Cox and Charles Duncan for their considerable contribution in bringing this series of study books to publication. We extend our grateful thanks for their unstinting patience and support throughout this process.

The authors and publisher would also like to thank the following for providing pictures for the book:

Bostick
Brady Corporation Ltd
Henkel Ltd
HSE
Ideal Industries Ltd
Kewtech
Maplin Electronics
Martindale Electric
Megger
MK Electric
Testmate

A special thank you to the ECA for providing the examples of Electrical Certificates for this book.

Every effort has been made to contact the copyright holders.

This book is endorsed by:

Representing the best in electrical engineering and building services

Founded in 1901, the Electrical Contractors' Association (ECA) is the UK's leading trade association representing the interests of contractors who design, install, inspect, test and maintain electrical and electronic equipment and services.

www.eca.co.uk

Study Guide

This study book has been written and compiled to help you gain the maximum benefit from the material contained in it. You will find prompts for various activities all the way through the study book. These are designed to help you ensure you have understood the subject and keep you involved with the material.

Where you see 'Sid' as you work through the study book, he is there to help you and the activity 'Sid' is undertaking will indicate what it is you are expected to do next.

Task

Familiarize yourself with the requirements for voltage detection instruments given in HSE Guidance Note GS 38 before you continue with this chapter.

Task A 'Task' is an activity that may take you away from the book to do further research either from other material or to complete a practical task. For these tasks you are given the opportunity to ask colleagues at work or your tutor at college questions about practical aspects of the subject. There are also tasks where you may be required to use manufacturers' catalogues to look up your answer. These are all important and will help your understanding of the subject.

Try this

State the:

1 Two test methods that are used to test for polarity

_____

_____

2 Effect on the operation of a circuit breaker connected in the neutral conductor only in the event of a fault to earth.

_____

_____

Try this A 'Try this' is an opportunity for you to complete an exercise based on what you have just read, or to complete a mathematical problem based on one that has been shown as an example.

Remember

The Electricity at Work Regulations place a responsibility on us to ensure that we do not leave an installation in an unsafe condition.

Remember A 'Remember' box highlights key information or helpful hints.

RECAP & SELF ASSESSMENT

Circle the correct answers.

1 Which of the following test instruments would be used to identify a short circuit between live conductors and earth?

   a.   approved voltage indicator
   b.   low resistance ohmmeter
   c.   insulation resistance ohmmeter
   d.   earth fault loop impedance tester

2 The document used to record that a repaired circuit is safe to put into service is a:

   a.   repair record sheet
   b.   Schedule of Test Results
   c.   Electrical Installation Condition Report
   d.   Minor Electrical Installation Works Certificate

Recap & Self Assessment At the beginning of all the chapters, except the first, you will be asked questions to recap what you learned in the previous chapter. At the end of each chapter you will find multichoice questions to test your knowledge of the chapter you have just completed.

Note

More detailed information on testing continuity of ring final circuits can be found in the *Inspection Testing and Commissioning* study book in this series.

Note 'Notes' provide you with useful information and points of reference for further information and material.

This study book has been divided into Parts, each of which may be suitable as one lesson in the classroom situation. If you are using the study book for self tuition then try to limit yourself to between 1 hour and 2 hours before you take a break. Try to end each lesson or self study session on a Task, Try this or the Self Assessment Questions.

When you resume your study go over this same piece of work before you start a new topic.

Where answers have to be calculated you will find the answers to the questions at the back of this book, but before you look at them check that you have read and understood the question and written the answer you intended to. All of your working out should be shown.

At the back of the book you will also find a glossary of terms which have been used in the book.

A 'progress check' at the end of Chapter 3, and an 'end test' covering all the material in this book, are included so that you can assess your progress.

There may be occasions where topics are repeated in more than one book. This is required by the scheme as each unit must stand alone and can be undertaken in any order. It can be particularly noticeable in health and safety related topics. Where this occurs, read the material through to ensure that you know and understand it and attempt any questions contained in the relevant section.

You may need to have available for reference current copies of legislation and guidance material mentioned in this book. Read the appropriate sections of these documents and remember to be on the lookout for any amendments or updates to them.

Your safety is of paramount importance. You are expected to adhere at all times to current regulations, recommendations and guidelines for health and safety.

Unit Seven

Fault Finding and Diagnosis

Material contained in this unit covers the knowledge requirement for C&G Unit No. 2357-307 (ELTK 07), and the EAL Unit QELTK3/007.

Fault finding and diagnosis considers the principles, practices and legislation for diagnosing and correcting electrical faults in electrotechnical systems and equipment in buildings, structures and the environment. It considers the reporting and recording of electrical faults, the preparatory work prior to fault diagnosis and the procedures and techniques for diagnosing and correcting electrical faults.

You could find it useful to look in a library or online for copies of the legislation and guidance material mentioned in this unit. Read the appropriate sections and remember to be on the lookout for any amendments or updates to them. You will also need to have access to manufacturers' catalogues for wiring systems, tools and fixings.

Before you undertake this unit read through the study guide on page viii. If you follow the guide it will enable you to gain the maximum benefit from the material contained in this unit.

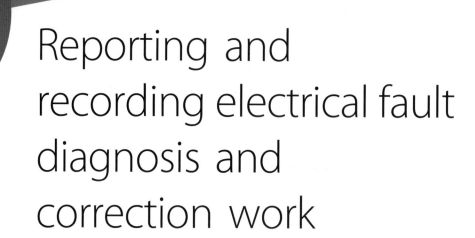

1

Reporting and recording electrical fault diagnosis and correction work

This study book considers the principles and practices of fault diagnosis, location and rectification.

LEARNING OBJECTIVES

On completion of this chapter you should be able to:

- State the procedures for reporting and recording information on electrical fault diagnosis and correction work

- State the procedures for informing relevant persons about information on electrical fault diagnosis, correction work and the completion of relevant documentation

- Explain why it is important to provide relevant persons with information on fault diagnosis and correction work clearly, courteously and accurately.

Part 1 Reporting and recording

This chapter considers the requirements for the reporting and recording of information related to fault diagnosis and corrective work. We shall start with a general overview of fault finding.

During our work activities we are inevitably going to come across equipment, circuits or cables which have faults. In some cases the client may report a fault with a circuit or piece of equipment which they want repaired. In general, the customer reports a problem when a circuit or piece of equipment fails to operate correctly or when it ceases to operate at all. It is important to consider the difference between the:

● Faults we identify on the electrical installation
● Faults referred by the customer.

We may identify faults on the electrical installation when carrying out tests during the initial verification. These will need to be corrected before continuing with the testing and to do this we will have to locate the fault and repair it. Any faults identified during this process will affect the safety of the installation if not corrected. Some of these faults would not be apparent to the client during the normal operation of the installation.

Note
The process of inspecting and testing electrical installations and the recording of the results has been detailed in the *Inspection, Testing and Commissioning* study book in this series.

For example, a circuit with an unacceptable earth fault loop impedance (Z_s) may only become apparent to the user when it is too late and

someone has suffered an electric shock. The purpose of the initial verification is to ensure that no such hidden dangers exist.

Remember
The Electricity at Work Regulations place a responsibility on us to ensure that we do not leave an installation in an unsafe condition.

The process of fault location requires some logical thought and is one which allows us to investigate the fault in the simplest way. Fetching a double extension ladder, climbing to a high outside light, replacing the lamp, only to find that someone has switched the circuit off is a typical example of beginning without a logical approach – checking the simple things first.

Figure 1.1 *Check first!*

Dismantling a domestic cooker to test the elements because the oven is not working, only to find the customer has switched the oven to timer is another typical example. When faced with a fault we need to approach it logically and methodically, starting with the obvious and progressing through one stage at a time.

Consideration should also be given to factors such as cost, resources, safety (of both personnel and the installation) and minimizing the amount of time the installation or circuit is switched off in order to avoid disruption.

Recording information

The process for recording and reporting information when dealing with electrical faults will vary from one company to another. First it is essential to gather all the relevant information relating to the reported fault. To do this, most companies will have a fault record which will need to be completed for each fault reported to them.

Example: A customer has telephoned an electrical contractor regarding a fault on their electrical installation. The record of this is shown in Figure 1.2 and we can see that the date and client details are relatively straightforward. The reported fault is identified as the client expressed it: 'upstairs lights flickering'. The other relevant details which are recorded are as a result of further questioning by the contractor. This established that it was only one light which was affected. The advice given to the client during the discussion was designed to minimize any danger until the fault could be rectified. The date actioned column records that the job has been issued to an electrician for investigation in the morning on an agreed date.

Some preliminary investigation is always beneficial and should enable the contractor to determine that a fault actually exists and offer advice to the customer.

JD Installations Ltd		Fault Notification Record				No. 00135
Date received	Client details	Reported fault	Symptoms	Relevant details	Advice given	Date actioned
--/07/20--	Mrs. J Smeath, 52 Dale Road, Whetherington, WS2 5HR	Upstairs lights flickering	When the main bedroom light is turned on the light flickers	Other upstairs lights operate OK. Only the main bedroom light affected	Avoid use of the light and use side lights until fault is rectified	--/07/20-- Sent to J. Douglas for action on --/07/20—am

Figure 1.2 *Typical report notification record*

Reporting on remedial action

Where a fault on the electrical installation is investigated, the remedial work will need to be certificated to confirm the installation is safe for use and complies with the requirements of BS 7671. This will normally be carried out using a Minor Electrical Installation Works Certificate (MEIWC).

MINOR ELECTRICAL INSTALLATION WORKS CERTIFICATE FOR UP TO THREE CIRCUITS ⚙ECA

(REQUIREMENTS FOR ELECTRICAL INSTALLATIONS BS 7671 (IET WIRING REGULATIONS))
To be used only for minor electrical work which does not include the provision of a new circuit

Certificate number: MW001235   Member number: [ ]

PART 1: DESCRIPTION OF MINOR WORKS

1. Description of the minor works Rectify fault on upstairs lights; replace faulty main bedroom ceiling rose, damaged by arcing at loose terminal

2. Location / address 52 Dale Road, Whetherington, WS2 5HR, Main Bedroom first floor front

3. Date minor works completed -- July 20 --

4. Details of departures, if any, from BS 7671:2008

NONE

PART 2: INSTALLATION DETAILS

1. System earthing arrangement   TN-C-S   TN-S ✓   TT
2. Method of fault protection ......ADS......
3. Protective device for the modified circuit   1 Type BS 3036   Rating 6   A
                                                   2 Type ....   Rating ....   A
                                                   3 Type ....   Rating ....   A

Comments on existing installation, including adequacy of earthing and bonding arrangements (see 132.16)
The electrical installation is over 40 yrs old and there are no cpcs for the lighting circuits. Inspection shows there are no Class 1 metallic
accessories installed. The client has been advised and improvement has been recommended.
Main protective bonding is in place to gas and water installation pipework (6 mm²) and addition 5yrs ago.

PART 3: ESSENTIAL TESTS

CCT No	Circuit description	Continuity ✓ or Ω (fill 1 col. only)		Insulation resistance MΩ			Polarity	Zs	RCD (if applicable)	
		R₁ + R₂	R₂	Line / neutral	Line / earth	Neutral / earth	✓	Ω	1 x IΔn	5 x IΔn
1	Upstairs lighting	N/A		198	198	198	✓	N/A	N/A	N/A
2										
3										

PART 4: DECLARATION

I / We CERTIFY that the works do not impair the safety of the existing installation, that the said works have been designed, constructed, inspected and tested in accordance with BS 7671:2008 (IET Wiring Regulations), amended to .... (date) and that the said works, to the best of my / our knowledge and belief, at the time of my / our inspection complied with BS 7671 except as detailed in Part 1 above.

Contractor's name: James Douglas   Signature: James Douglas
For and on behalf of: JD Installations Ltd   Position: Proprietor
Address: Unit 6, Gambols Estate, Neartown, Cutcounty, NT5 8LR   Date: -- July 20 --

C-MW3-ECA REV V1 Aug 2011

Figure 1.3 *MEIWC for the rectification of the fault*

In our example, the electrical installation was over 40 years old with no circuit protective conductor (cpc) installed for the lighting circuit and

so it was not compliant with the current requirements of BS 7671. The electrician inspected the installation and determined that there were no Class I metallic accessories installed and therefore not an immediate risk of electric shock to the user of the lighting circuit.

As the ceiling rose had been damaged by arcing at a loose terminal and represented an immediate fire risk, the electrician replaced the ceiling rose to remove the immediate danger. The client was advised of the situation and the fact that remedial work was required which was then recorded on the MEIWC.

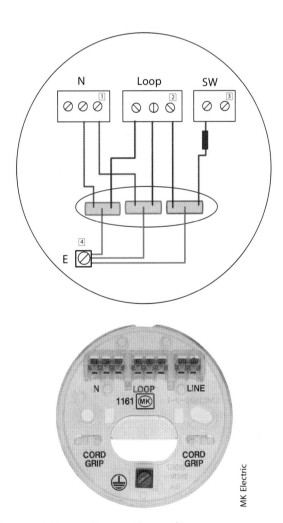

Figure 1.4 *Replacing the ceiling rose*

MK Electric

When dealing with faults on electrical installations it is important to rectify an immediate danger wherever possible without delay. A check list for the action includes identifying if:

- There is an immediate danger to the user of the installation
- There are any other departures from BS 7671
- These other departures present an immediate risk to the user.

Any departures which represent an immediate danger to the user should be rectified as a matter of urgency. Those departures which do not represent an immediate danger must be advised to the client and recorded on the certificate.

In our example the electrician determined there was an immediate risk from fire. The circuit did not meet the current requirements of BS 7671, but the inspection established that there was no immediate danger to the user due to the lack of cpc and so carried out the repair.

Where faults involve equipment and machines the recording process is slightly different. The fault is often reported simply as an item of equipment which is not working, or not working as it should.

In such cases we need to consider the information relating to the equipment such as the maintenance records. These should indicate the previous performance history of the equipment.

When reporting on fault finding, some companies use the maintenance records whilst others have a separate document for recording the fault and remedial action. The repairing of faults to motors and the like does form part of their maintenance and so whichever format is used, the information should be kept with the maintenance records.

Charles Machinists Ltd		Repair record				No. 00173
Date	Equipment Ref No	Fault	Symptoms	Cause	Material	Re-commissioned
--/07/20--	Workshop heater WH1	No heat and fan boost not operating	Operating OK and then stopped working	Traced to thermal cut-out fused	Replacement thermal cut-out TK4 R120	--/07/20—am
Name: James Douglas			Signature: *James Douglas*			Date: --/07/20--

Figure 1.5 *Typical repair record sheet*

Task

A client has contacted your company to report a fault on their car park lighting as only one of the lights is working. Compile a list of questions you would ask the client before setting out for the site.

_____

_____

_____

_____

_____

_____

_____

_____

_____

_____

_____

_____

_____

Part 2 Providing information

The client, customer and users of the installation and equipment need to be kept informed on the activities, requirements and actions necessary during the fault diagnosis and repair process.

Information

The need to both consult and keep the client informed is an essential part of the fault diagnosis and repair for a number of reasons, including:

● Information to be obtained to help in the diagnosis

● The effects of the diagnosis on the normal activities of the client or user
● The time scale for the diagnosis and repair
● The viability of repairs (particularly to equipment)
● The cost implications of repair or replacement
● Progress.

It is very difficult to predict the time required to locate and repair faults. Fault finding, by its very nature, requires a degree of investigation and the more details we can establish about the cause or symptoms the better. This helps to focus the area of investigation and often reduces the time taken to identify and locate the fault.

Of course, once the fault is located there is the question of repair or replacement. In either event there will be a requirement for materials or parts and these will need to be sourced. For common items these are often available off the shelf and therefore do not add significantly to the time for repair. Less common items frequently have a delay period awaiting delivery and this will subsequently delay the repair process. In extreme cases parts may need to be manufactured and this can add a considerable delay to the process.

In the first instance the client will need to know when we are going to attend to investigate the problem. The response time will be dependent upon:

● The availability of labour
● The urgency of the problem.

This needs to be dealt with courteously and politely as, for most people, when something is not working the repair is urgent to them. There are some obvious occasions where this may not be the case and the repair can be dealt with later. On other occasions the situation is urgent and requires our urgent attention.

It is also important to discuss the actual process with the client, including such things as the need to isolate circuits or equipment, where isolation will be carried out, what will be affected and when would be a convenient time to do this.

As we mentioned earlier there will also be a need to discuss the remedial action as this will have both time and financial implications for the client. It is important to give the client any technical detail in layman's terms, ensuring this is understood without talking down to the client. It is also helpful to the client to offer a qualified engineering opinion on the suitable options for

remedial action. For example; is a repair worthwhile or is the condition of the equipment so poor that a repair is not a viable option.

Figure 1.6 *Equipment in poor condition*

Documents

The documents we need to consider are in two main categories.

Those provided by the client

This will include any maintenance logs, manufacturers' instructions and company record documents for faults and repairs. We need to establish which of these documents are available from the client and the significance any missing information will have on our activities.

Where the problem relates to the electrical installation the Electrical Installation Certificate, Minor Electrical Installation Works Certificate and any Electrical Installation Condition Reports (EICRs) (formerly Periodic Inspection Reports (PIRs)) should be available. These are required to identify the isolation and switching arrangements, the circuit identification, protective devices and so on. These documents will also provide evidence of the condition of the electrical installation over time and help to identify any deterioration that may have taken place.

In many simple installations, such as dwellings, there is often no such information available, so we have to rely on any information relating to the fault and installation which can be given to us by the user.

Remember

The information on the isolation, switching and control of the installation is essential for fault finding to be carried out safely.

Those documents we provided to the client

These are the documents issued on completion of the fault finding and repair process. In the case of a client who keeps maintenance records these results are often recorded on their own documents which they will provide to us. In other instances we have to provide the client with documentary records of the work carried out. We have seen samples of these documents in Figures 1.2 and 1.3 in this chapter.

The MIEWC is appropriate where a repair is carried out to a circuit. If the circuit has to be re-wired then an Electrical Installation Certificate (EIC) will have to be issued.

It is important to advise the client of the nature and purpose of the documentation we issue and there is some guidance for the recipient provided with the standard forms in BS 7671. However it is important that the purpose and retention of these forms is explained to the client.

Where the fault relates to a machine or item of equipment then a form such as that in Figure 1.3 may be suitable for the client. This should be accompanied by the details and results of any test carried out. Where the circuit supplying the equipment was also tested to confirm that it was safe for use, these results may be in the form of the MEIWC.

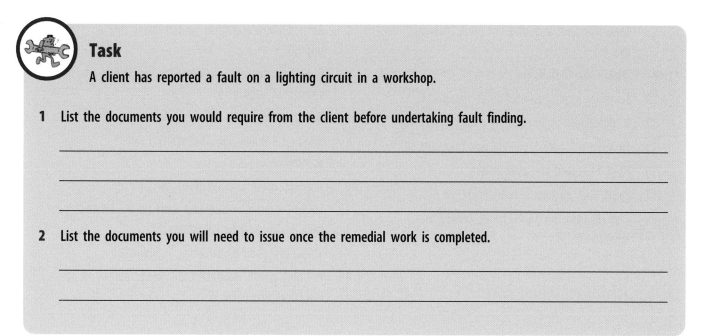

Task

A client has reported a fault on a lighting circuit in a workshop.

1 List the documents you would require from the client before undertaking fault finding.

_____

_____

_____

2 List the documents you will need to issue once the remedial work is completed.

_____

_____

Try this: Crossword

Across

5   A prime consideration when fault finding (6)

6   Once the fault is found we must identify this (5)

8   Where the fault is (8)

10 Financial implications of fault finding (4)

11 A logical one is needed for fault finding (7)

13 To ensure 5 across you do not take this (4)

15 Used to record the events (9)

17 Fault ... is putting it right (10)

18 General term for tools, parts, labour and equipment (9)

Down

1   Not having the spare parts may cause this (5)

2   Both doctors and fault finders carry this out (9)

3   Providing this ensures disconnection from all sources of supply (9)

4   Put it right (6)

7   Without 3 down one of these is a real possibility (5)

9   A cover against faulty parts (8)

12 Could be good or bad and with an 'er' on the end may be applied to your hair (9)

14 These are used to arrive at 2 down (8)

16 In short this is produced following repairs on an electrical circuit (5)

Congratulations, you have now completed the first chapter of this study book. Complete the self assessment questions before continuing to the next chapter.

SELF ASSESSMENT

Circle the correct answers.

1   One of the documents which should be requested from the client to prepare for fault finding on an injection moulding machine is the:

   a.   Minor Electrical Installation Works Certificate

   b.   Electrical Installation Certificate

   c.   maintenance schedule

   d.   manufacturer's warranty

2   One of the key pieces of information to be given to the client before carrying out fault finding is the:

   a.   exact time the work will take

   b.   isolation of equipment

   c.   total cost of the work

   d.   materials required

3   Communication with the client regarding fault diagnosis should always be clear, courteous and:

   a.   accurate

   b.   written

   c.   verbal

   d.   technical

4   A Minor Electrical Installation Works Certificate would be issued following the fault correction work on a:

   a.   vacuum forming machine

   b.   single phase pillar drill

   c.   three phase motor

   d.   lighting circuit

5   There is no information available on an item of equipment which has been reported as faulty. The most likely effect on the fault finding process is that it will:

   a.   cost more

   b.   take longer

   c.   require replacement parts

   d.   have to be done out of hours

2

Safe working procedures prior to fault diagnosis

RECAP

Before you start work on this chapter, complete the exercise below to ensure that you remember what you learned earlier.

- Faults found on the electrical installation during the _____ verification must be _____ before continuing with the _____.

- The process of fault location requires a _____ process which allows investigation of a fault in the most _____ way.

- When rectifying faults consideration should be given to such factors as cost, _____, safety and _____ the time the installation or circuit is _____ to avoid disruption.

- The remedial work associated with repairing a fault on a _____ will be using a Minor Electrical _____ Works _____.

- When investigating faults on electrical installations it is important to _____ an immediate _____ wherever possible without _____.

- The client, customer and _____ of the installation and equipment need to be kept _____ on the requirements and _____ necessary during the fault _____ and _____ process.

- It is important to give the client any technical detail in _____ terms and ensure that this is _____.

Carrying out and ensuring safe isolation is an essential part of an electrician's work. For this reason the requirements for safe isolation appear in a number of the units of the national occupational standard. This chapter considers the requirements for safe isolation in relation to fault finding and rectification work. In doing so it includes additional information to that contained in the other study books.

LEARNING OBJECTIVES

On completion of this chapter you should be able to:

- Specify safe working procedures that should be adopted for completion of fault diagnosis and correction work, including:

  - Effective communication with others in the work area

  - Use of barriers

  - Positioning of notices

  - Safe isolation.

- Specify the correct procedure for completing the safe isolation of an electrical circuit with regard to:

  - Assessment of safe working practices

  - Correct identification of circuits to be isolated

  - The selection of suitable points of isolation

  - The selection of correct test and proving instruments in accordance with relevant industry guidance and standards

  - The use of correct testing methods

  - The selection of locking devices for securing isolation

  - The use of correct warning notices

  - The correct sequence for isolating circuits.

- State the implications of carrying out safe isolation to:

  - Other personnel

  - Customers/clients

  - Public

  - Building systems (loss of supply).

- State the implications of not carrying out safe isolation to:

  - Self

  - Other personnel

  - Customers/clients

  - Public

  - Building systems (presence of supply).

- Identify all Health and Safety requirements which apply when diagnosing and correcting electrical faults in electrotechnical systems and equipment, including those which cover:

  - Working in accordance with risk assessments/permits to work/method statements

  - Safe use of tools and equipment

  - Safe and correct use of measuring instruments

  - Provision and use of personal protective equipment (PPE)

  - Reporting of unsafe situations.

Part 1 Statutory requirements

This chapter considers the requirements of the Electricity at Work Regulations (EWR), for safe isolation of electrical circuits and installations, which enable electrical fault finding and repair work to be carried out safely.

Whilst working through this chapter you will need to refer to the Memorandum of Guidance to the Electricity at Work Regulations and the Health and Safety Executive's HSE Guidance Note GS 38, Electrical test equipment for use by electricians.

Note

The Memorandum of Guidance to the Electricity at Work Regulations and the Health and Safety Guidance Note GS 38, Electrical test equipment for use by electricians, are both available as free downloads from **www.hse.gov.uk**.

Legal requirements

The Electricity at Work Regulations is the principal legislation relating to electrical activities in the workplace. It places duties on all those involved in terms of the work activities and the actions of the individuals.

The two main requirements we shall consider are those relating to the construction of the installation and the safety and actions of the operatives.

With regards to the installation construction, all electrical installations are to be constructed so that they are safe for use and can be maintained, inspected and tested and altered safely.

The requirements for the operatives are that they:

● Are competent to carry out the work they undertake

● Ensure the safety of themselves and others in so far as is within their control.

Competence: this study book considers the competence requirements specifically related to fault finding and subsequent corrective work. This calls for the knowledge and understanding of the requirements of the electrical installation and equipment which is to be worked on.

Safety: this is essential during the fault finding and repair of electrical installations and equipment. Regulation 4(3) of EWR requires that *'work on or near an electrical system shall be carried out in such a manner as not to give rise, so far as is reasonably practicable, to danger'*.

This means the safety of the operative and anyone in the vicinity or within the premises must be safeguarded. The requirements for safety can be achieved by following appropriate procedures for fault finding as detailed in this study book.

One of the main risks associated with working on electrical installations and equipment is that of electric shock. Regulation 14 of EWR refers to working on or near live conductors and is one of the regulations that are classed as 'reasonably

practicable'. This allows a risk assessment to be undertaken for this activity.

Working on or near live conductors should only be undertaken where:

● It is unreasonable under all circumstances for the equipment to be dead

● It is reasonable for the work to take place on or near live conductors

● Suitable procedures have been taken to prevent injury.

When undertaking fault finding the only reason for the installation, circuit or equipment to be live is when 'live testing' is necessary. The majority of fault finding is carried out with the installation, circuit or equipment isolated from the supply.

When undertaking fault finding and repair it is important to ensure that the part of the installation we are going to work on is safely isolated from the supply.

Note

It is never acceptable to work on live equipment for the convenience of the operative or as a result of pressure from the client. It is also not acceptable as an option to 'save time'. There is no reason that any process can be valued above the cost of life so unless exceptional conditions exist there should be no reason for carrying out live working in a consumer's installation.

Task

1   State two statutory responsibilities which apply to you when undertaking fault finding on electrical equipment.

_____

_____

_____

2   You are asked by the client to keep equipment in operation whilst carrying out fault finding on an installation. State how you would explain, giving the reasons why, this is not possible if you are to ensure the safety of all concerned.

_____

_____

_____

_____

_____

_____

Part 2 The need for safe isolation

Safe isolation does not simply mean making sure that the supply is switched off; it also includes making sure that it is not inadvertently re-energized.

As we have seen many of the fault finding activities we undertake will require safe isolation to enable us to work safely. This action will have implications for building systems, other people and ourselves. Similarly, failure to carry out safe isolation will also have its implications and perhaps it would be as well to consider these first.

Figure 2.1 *Safe isolation is required*

Remember

Fault finding and repair is undertaken on existing installations which are energized and in service. Safe isolation is an essential part of this activity.

Failure to safely isolate

The most obvious implication of the failure to safely isolate when working on electrical installations and equipment is the risk of electric shock to us and others.

The effect of electric current on the bodies of humans and animals is well recorded and the values quoted here should be taken as general guidance only. The average human body resistance is considered to be in the region of 1 KΩ (1000 Ω) and when subjected to a voltage of 230 V the current would be in the region of 0.23 A (230 mA).

A current flowing across the chest of a person in the region of 50 mA (0.05 Amperes), or more, is enough to produce ventricular fibrillation of the heart which may result in death. (Ventricular fibrillation is where the heart rhythm is disrupted and results in irregular fluttering rather than the beat required to circulate blood around the body.)

When working on the electrical installation or equipment we often have to expose live parts, which if not safely isolated, pose a serious risk of electric shock. Other people and livestock within the vicinity of our work will also be able to access these live parts and may not have sufficient knowledge and understanding to avoid the dangers involved.

Where electrical work is carried out in public areas this risk is further increased as the installation and equipment may be accessed by anyone – adults, children and animals – and the failure to safely isolate presents a very real danger.

Task

Using the information available on the Internet

1 State:

    a When ventricular fibrillation occurs, which area of the heart is affected?

    _____

    b What action would be required to help the casualty?

    _____

    c How the action in (b) above helps to save life.

    _____

a)

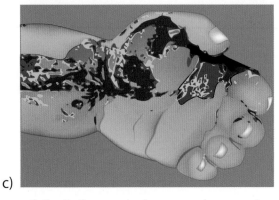

b)

c)

Figure 2.2 *Failure to isolate may have serious consequences*

Electric shock also carries the danger of electrical burns which occur at the entry and exit points of the contact and within the body along the path taken by the current. These

burns can be severe and whilst the casualty may survive the electric shock, the damage—some of which may be irreparable, can be considerable.

Whilst an electric shock may not be severe or fatal there is a risk of further injury as a result of it. Injuries which may occur include falls from steps and ladders, injury from machinery and vehicles, all of which are caused by the involuntary reaction when a person receives an electric shock.

To ensure these dangers are removed safe isolation of the installation, circuit or equipment is essential.

Failure to isolate can also affect the building and structure. Failure to isolate introduces a risk of arcing where live parts are exposed. This may occur between live parts at different potentials (line to neutral and between line conductors) and between live conductors and earth.

When an arc is produced, electrical energy is converted into heat energy and the level of discharge energy results in molten conductor material being present in the arc. This presents a real risk of fire and the automatic operation of the protective device isolating the circuit will not extinguish a fire started in this way.

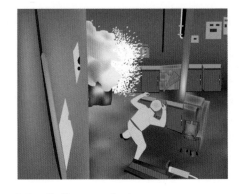

Figure 2.3 *Failure to isolate*

> ### Note
> The battery used to power a wristwatch can produce enough energy in a spark to cause an explosion or fire in a flammable or explosive atmosphere.

Figure 2.4 *Services may be affected*

In the case of a fault, including one which does not result in an electric shock or fire, the supply should be disconnected automatically. In this case the circuit or installation may be disconnected unexpectedly. The building systems may be switched off resulting in loss of data, failure of heating or ventilation, lighting and power.

In some instances we may also lose the building and life protection systems such as fire alarm systems, sprinklers, smoke vents, firefighters' lifts and the like. It may also result in the loss of lighting and ventilation to internal areas that have no natural light or ventilation.

This can result in considerable expense to the client and damage to the electrical equipment and buildings. The loss of lighting can cause other dangers to persons within the building resulting in trips, falls and injuries from machinery and equipment.

Effects of safe isolation

It is important to remember that during the fault finding process not only do we need to consider safe isolation, we also have to determine the most appropriate place to isolate the equipment being worked on. Carrying out safe isolation is essential to safeguard against the dangers we have identified above but there are a number of implications which must be considered.

Before we isolate we need to consider the effect this will have on persons, the building and the equipment and services. We need to determine the extent of the installation which needs to be isolated to carry out our work safely. There are certain activities which will need the whole installation to be isolated. However, the majority of fault finding involves a particular circuit, item of equipment and possibly a system such as the air conditioning. In this case the safe isolation of one or more circuits or items of equipment may be all that is necessary.

Figure 2.5 *Safe isolation will affect others*

Carrying out isolation has implications for the users of the installation as electrical equipment will not be available for use. This means that the timing and duration of the work must be carefully considered and discussed with the user to minimize disruption.

In the case of complex installations, permits to work systems are used to control and authorize the work to be carried out. The control of activities has to be coordinated and clear instructions are necessary to ensure the safety of everyone within the installation. The extent of work carried out and the duration it is clearly identified and method statements are required to ensure safety.

> **Note**
>
> The use of method statements and permits to work are covered in more detail in the *Organizing and Managing the Work Environment* study book in this series.

We will need to consider providing task lighting and power for our work depending on the location, as it may not be well lit to enable work to be carried out safely. Where fault finding involves lighting circuits there will be no supply available so it may be necessary to arrange temporary lighting system for the client.

If any safety services or alarms are affected by the need to isolate, we need to consider the consequences of this. For example, burglar and other alarms may be linked to the emergency services and the loss of supply may result in a response visit.

Other building services which may be affected include door access systems, bar code readers and tills, security cameras and public address systems.

Figure 2.6 *Safe isolation may affect safety services and alarms*

Preparation for fault finding

As much of the work is likely to be carried out in areas which are occupied by the client, staff, visitors or the general public, we need to make certain appropriate precautions are taken to safeguard both us and anyone else in the vicinity of our work.

It is common for fault finding to be carried out in a pressured environment with the client and operatives eager for:

- The fault to be repaired and normal activity to be resumed
- No additional interference with the operation of the business.

This is particularly noticeable in areas where production is being affected as this may have a direct effect on the business profit and the workers' wages.

It is important to ensure that the fault finding is carried out effectively and efficiently without any compromise to the safety of everyone.

The first task is to discuss with the client the actions to be taken and the means to keep disruption to a minimum. This includes permission to isolate circuits or equipment appropriate to the task. It may also involve the requirement for you to provide method statements and for permits to work to be issued.

In some cases it may not be possible to carry out the work during normal working hours. Arrangements would then need to be made with the client to obtain access and for a member of their staff to attend, for security or access purposes, whilst the work is carried out.

Having agreed these arrangements with the client the next job is to inform those in the area where the work is to be carried out. By explaining the activities that will be going on and what is involved this will:

● Remove speculation and numerous enquiries whilst you are working
● Make people aware of the process and the precautions in place for their safety
● Enforce the need for care and tolerance on their part.

This information is reinforced by the use of appropriate measures to ensure a safe working environment. This will include the use of:

Barriers: these are used to cordon off the work area to create a safe environment to work in and prevent unauthorized personnel entering the area. The use of proprietary barriers or warning tapes is generally the best method of achieving this.

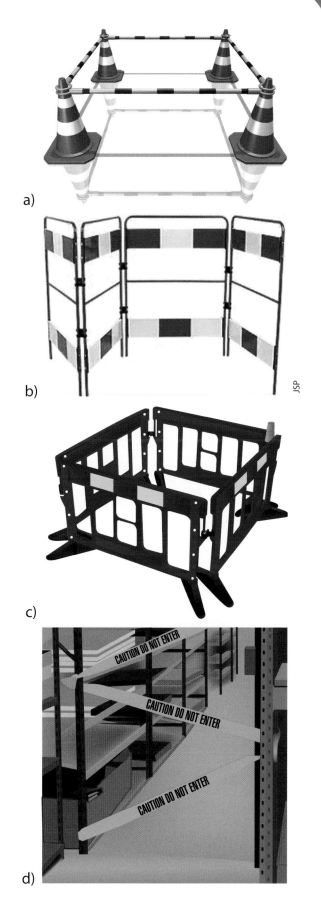

a)

b)

c)

d)

Figure 2.7 *Warning barriers and tapes*

Suitable signs: these will need to be positioned in a number of places whilst the work is being undertaken. The use of self supporting warning signs at the access points to the work area reminds people of the activity. In areas where access is open to a large number of staff or members of the public, placing warnings at the entrance to the building, the specific floor or approaches to the work area will be necessary.

Notices: these will be necessary at the points of isolation and at the equipment on which work is carried out. This informs that isolation has been undertaken for a specific reason and that it is under the control of a competent person. The signs on or adjacent to equipment inform people why the equipment is not functioning and prevents unauthorized interference.

Figure 2.8 *Warning signs*

Figure 2.9 *Warning notices*

Safetyshop, A Division of Brady Corporation Ltd.

Task

A fault has been reported which has caused one machine in a busy workshop to stop working. You have been asked to investigate and repair the problem.

1 State what must be agreed with the client before any work is undertaken.

_____

_____

2 Explain three actions that can be taken to ensure safety of the employees whilst the work is carried out.

_____

_____

_____

_____

Part 3 Safe isolation procedures

Safe isolation is one of the key factors in ensuring safety during fault finding. As discussed earlier, equipment or circuits being investigated should be isolated whilst the work is carried out. The **only** exception is where live testing is necessary.

The key to safety is to follow the correct procedure(s) throughout the isolation process. Let's consider a situation where we are going to carry out fault finding on a lighting circuit in a shop. Before we can begin fault finding we must safely isolate the lighting circuit from the supply. This particular circuit is protected by a BS EN 60898 type B circuit breaker (cb) in the distribution board located at the origin of the installation.

Figure 2.10 *Safe isolation includes securing the control*

Before we can begin we must first obtain permission to isolate the circuit from the person responsible for the electrical installation (the duty holder). As we are going to be isolating the lighting circuit the duty holder must ensure that the safety of persons and the operation of the business is not going to be compromised.

To do this the area to be affected and the duration of the isolation should be explained to the duty holder to help in making the decision.

Having been given permission to isolate the circuit we must correctly identify the relevant one within the distribution board. Where there are a number of lighting circuits it is important that we isolate the right one. Providing the distribution board has been correctly labelled and the appropriate circuit charts are available this should be relatively straightforward.

Test instruments which allow the identification of a circuit before it is isolated are available. These operate using:

- A transmitter unit connected to the circuit which then sends a signal along the circuit conductors
- A detector unit is used to sense the signal at the distribution board to identify the fuse or circuit breaker.

The sensitivity of the receiver unit can be adjusted to give a clear and reliable indication of the circuit to be isolated.

This is quite straightforward where the circuit includes a socket outlet, as the transmitter unit can be readily plugged into the circuit. However, where the circuit does not include a suitable socket, a connection needs to be made to live parts.

This introduces a higher risk to the operative as the circuit will not be isolated; the device is being used to identify a circuit so isolation to access the live parts is not an option. In such circumstances this operation will require two persons: one to operate the sensing unit and the other to make the connection to the circuit using test probes complying with GS 38. Extreme care is required when accessing the live terminals to make this connection and it should only be carried out by skilled persons using suitable equipment.

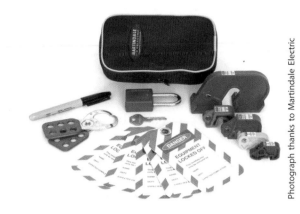

Photograph thanks to Martindale Electric

Figure 2.12 *Typical lock off kit*

Testmate

Figure 2.11 *Typical circuit identification or fuse finder instrument*

Having correctly identified the circuit, the circuit breaker is switched off, isolating the circuit from the supply.

An appropriate locking off device is then fitted to the cb to prevent the circuit from being unintentionally re-energized. There are a number of devices available for this task and they all perform the same function, preventing the operation of the cb. In most cases a separate padlock is inserted through the locking off device to prevent the unauthorized removal and secures the cb. A warning label should also be fitted to advise that the circuit should not be energized and that someone is working on the circuit.

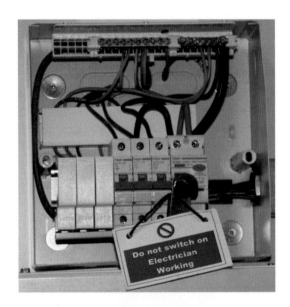

Figure 2.13 *Lock off in position*

Having safely isolated and locked off the circuit we must now confirm that the circuit is actually isolated from the supply. To do this we will need an approved voltage indicator (AVI) together with a proving unit. The term approved voltage indicator refers to a voltage indicator which meets all the requirements of Health and Safety Guidance GS 38.

Task

Familiarize yourself with the requirements for voltage detection instruments given in HSE Guidance Note GS 38 before you continue with this chapter.

Figure 2.14 *Typical proving unit and approved voltage indicator*

In this instance we are going to remove the cover from a luminaire on the circuit at a point close to the distribution board and confirm that it is actually isolated. The first step, having removed the cover, is to confirm that the approved voltage indicator is functioning using the proving unit. The output produced by the proving unit should cause all the light emitting diode (LED) indicators to light showing that they are all functioning correctly.

As this is a single phase lighting circuit we will need to confirm isolation by testing between:

● All line conductors and neutral (loop line terminal and switch line terminal operating the switch to confirm in both switch positions)

● All line conductors and earth (loop line terminal and switch line terminal)

● Neutral and earth.

And there should be no voltage present at any of these connections.

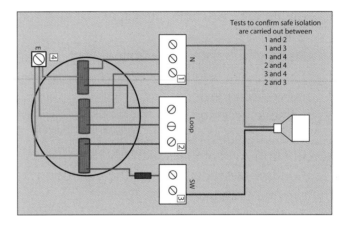

Figure 2.15 *Tests for safe isolation at luminaire*

Finally, we must confirm that the approved voltage indicator is still working. To do this we will use the proving unit again and all the LED indicators should light when the approved voltage indicator is tested. This process will have confirmed that the circuit we are going to work on is isolated from the supply.

It is always advisable, when removing accessories to carry out work, to further check at each accessory that isolation is achieved. It may be that a luminaire or accessory which appears to be on the circuit is actually supplied from elsewhere.

For example, it is not uncommon to find a downstairs light switch controlling an upstairs light as with two-way lighting on stairs. If the downstairs circuit is isolated it would be logical to assume that all the ground floor switches are isolated. Similarly, where two circuits supply the same area, such as the socket outlets in a workshop, then it is not always obvious which sockets are supplied on which circuit.

Summary procedure for circuit isolation

1   Seek permission to isolate
2   Identify the circuit to be isolated
3   Isolate by switching off circuit breaker or isolator
4   Fit locking device and warning label
5   Secure area around accessory to be removed (barriers)
6   Remove accessory
7   Select an approved voltage indicator (AVI)
8   Confirm AVI complies with GS 38
9   Confirm the operation of the AVI using a proving unit
10   Test between all live conductors

11   Is circuit dead? If not go back to 2
12   Test between all live conductors and earth
13   Is circuit dead? If not go back to 2
14   Confirm AVI is functioning using proving unit

There may be occasions when it is necessary to isolate a whole distribution board or a complete installation.

Remember

When isolating a number of circuits it is important to discuss with the duty holder the areas that will be isolated and determine any special requirements for the circuits which are to be isolated.

If we are to isolate a three phase distribution board, the process will be similar to that for isolating the single circuit. As there are three line conductors there will be more tests carried out to confirm isolation has been achieved. This same test sequence will be used if a three phase and neutral circuit is to be isolated and so it is worth considering the process here.

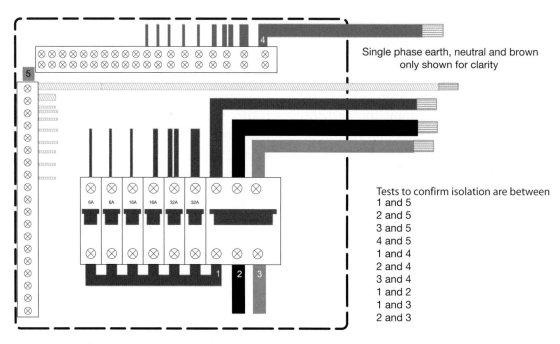

Single phase earth, neutral and brown only shown for clarity

Tests to confirm isolation are between
1 and 5
2 and 5
3 and 5
4 and 5
1 and 4
2 and 4
3 and 4
1 and 2
1 and 3
2 and 3

Figure 2.16 *Three phase points of test for isolation*

Summary procedure for distribution board isolation

1   Seek permission to isolate
2   Identify the distribution board to be isolated
3   Isolate by switching off main isolator
4   Fit locking device and warning label
5   Remove distribution board cover to access live terminals
6   Select an approved voltage indicator (AVI)
7   Confirm AVI complies with GS 38
8   Confirm the operation of the AVI using a proving unit
9   Test between all live conductors
10 Is circuit dead? If not go back to 2
11 Test between all live conductors and earth
12 Is circuit dead? If not go back to 2
13 Confirm AVI is functioning using proving unit

Note

A safe isolation flow chart is included at the end of this chapter for your reference. You can copy this and put it with your test equipment as an aide memoire.

Having determined the procedure for safe isolation we must also consider the location at which the isolation takes place. When working on equipment there is often a local isolator or control which may be used to isolate the equipment. First we need to establish whether the local control or isolator is suitable for use as an isolator.

Part 5 of BS 7671 provides some guidance on the devices which can be used to provide isolation in Table 53.4, Guidance on the selection of protective, isolation and switching devices.

From the devices listed in the table we can see that we cannot use semi-conductor devices, switching devices which do not comply with either BS EN 60669-2-4 or BS EN 60947-3 and contactors complying with BS EN 61095 to provide isolation. A number of the other devices have some conditions attached to their use and the design of the installation should be such that the installed devices are suitable.

The most appropriate place to isolate does depend on what the fault is and where it is likely to be. If a machine (or item of equipment) has stopped working but other equipment on the same circuit still functions then the fault is likely to be with the machine. If everything on the same circuit is not working then the circuit needs to be investigated, and so on.

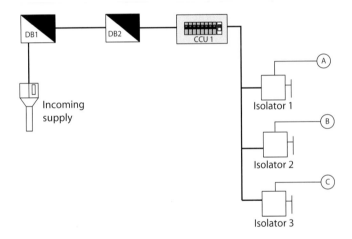

Figure 2.17 *Block diagram*

As we can see from the diagram in Figure 2.17 above, there are at least four points where we could isolate on the circuit shown, and the point which causes least inconvenience but provides safety for the work to be carried out, is the one to choose.

Task

Familiarize yourself with the content of Table 53.4 in BS 7671 before you continue with this chapter.

Remember

When we are proving the operation of the AVI we can use a known live supply or a proving unit. When isolating a distribution board it is possible to use the incoming supply to the isolator to prove the AVI is functioning before and after we test for isolation. This is not possible when isolating a single circuit and so a proving unit is essential for circuit isolation.

Remember

All items of test equipment, including those items issued on a personal basis, should be regularly inspected and where necessary, tested by a competent person.

Task

A three phase and neutral machine has developed a fault and is to be isolated from the supply to allow the fault to be investigated. The local isolator has been locked in the off position and the motor terminal cover removed. Describe, using the terminal block information shown in Figure 2.18, the procedure to be carried out to confirm the motor is isolated.

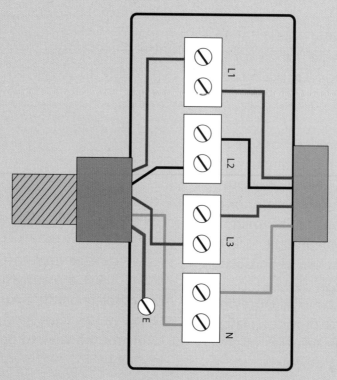

Figure 2.18 *Three phase and neutral machine terminals*

_____

_____

_____

_____

_____

_____

_____

_____

_____

Part 4 Health and safety requirements

Complying with the applicable legislation, standards and codes of practice is essential when carrying out fault finding on electrical installations.

The legislation which we must observe includes:

- The Health and Safety at Work (etc.) Act (1974) and
- The Electricity at Work Regulations (1989).

Both of these require the operatives to be competent to carry out the work they undertake.

A number of regulations and codes of practice have been introduced under the Health and Safety at Work (etc.) Act. Any of these regulations or codes of practice which are applicable to working on an electrical installation, including fault finding, or to the environment in which this work is undertaken, must be complied with.

BS 7671 and Institution of Engineering and Technology (IET) Guidance Note 3, Inspection and Testing, identify the need for competence and when carrying out testing, the requirements include the use of suitable and appropriate test instruments. This means the operative must be fully versed in the testing process and have suitable knowledge of the capability and limitations of the test equipment.

Remember

A competent person is defined by the Construction (Health, Safety & Welfare) Regulations 1996 as: 'Any person who carries out an activity shall possess such training, technical knowledge or experience as may be appropriate, or be supervised by such a person.'

Note

The statutory and health and safety requirements are considered in greater detail in the _Legislation_ and _Termination and Connection of Conductors_ study books of this series.

Tools, equipment and instruments

The tools and test equipment should be suitable for their intended use and for the conditions in which they are to be used. They need to be maintained and inspected periodically to ensure they remain safe for use. Records should be kept of the inspections carried out on the tools and equipment.

An employee is required by law to:

● Not interfere with tools, equipment, etc. provided for their health, safety and welfare

● Correctly use all work items provided in accordance with instructions and training given to them.

Test instruments are to be regularly calibrated and records of ongoing accuracy must be kept to confirm the instruments continue to give accurate test results.

Instruments should be treated with respect and it is important to remember that we rely on them being in good condition and to provide accurate readings. These results are used to identify and locate faults on the installation and equipment. Following repair they will also be used to confirm that the installation or equipment is safe to put back in service.

When testing with voltages in excess of 50 V ac the test leads should comply with the requirements of HSE Guidance Note GS 38, Electrical test equipment for use by electricians. Test leads are subject to wear and tear and they should be checked regularly to ensure they are in good condition and comply with GS 38.

Personal Protective Equipment (PPE)

Fault finding is no different to any other activity in respect to the requirements for PPE. Wherever PPE is provided for an activity it must be used and maintained in good condition. The PPE needs are generally based upon other people's activities within the area and the requirements of the location in which the work is carried out.

Safety glasses are one item of PPE equipment which should be used when carrying out testing. The danger of molten material (produced from arcing should an error be made) entering the eyes, is to be protected against at all times.

Remember

PPE is the final option in the safety actions for personnel. It is only used where the risk cannot be removed or dealt with in any other way.

Reporting of unsafe situations

Any unsafe situations identified during the fault finding process should be reported to the responsible person without delay. Any items which require immediate action and are within your control should be remedied.

Note

The reporting of unsafe situations does vary from one company to another and the requirements are covered in more depth in the study book *Termination and Connection of Conductors*.

Task

List the accident causes which occur when carrying out testing as identified in HSE Guidance GS 38.

Congratulations, you have now completed Chapter 2 so correctly complete the following self assessment questions before you carry on to the next chapter.

SELF ASSESSMENT

Circle the correct answers.

1 When carrying out maintenance on a machine, safe isolation is carried out at a local isolator. The effect of this action on other equipment on the same circuit will be the:

   a.   equipment will not function

   b.   equipment will function normally

   c.   use of the equipment will be restricted

   d.   performance of the equipment will be limited

2 The locking off of an isolator during fault finding on a circuit means that the circuit:

   a.   can be energized by the operative

   b.   can be readily operated by the client

   c.   cannot be inadvertently energized

   d.   cannot be tested

3 Whilst carrying out fault finding on a circuit it is necessary to remove accessories. Before an accessory is removed the first action by the electrician is to:

   a.   confirm the accessory is isolated

   b.   disconnect the conductors

   c.   ensure polarity is correct

   d.   notify the client

4 The maximum recommended rating of the fuse used to protect instrument test leads given in GS 38 is:

   a.   200 mA

   b.   300 mA

   c.   400 mA

   d.   500 mA

5 The requirement for a competent person carrying out fault finding is that they are familiar with the equipment and installation being worked on, and:

   a.   the test instruments to be used

   b.   only work on live equipment when told to do so

   c.   are careful when working on existing circuits

   d.   are supervised by a competent person

Safe isolation flow chart

```
                    ┌──────────────────────┐
                    │ Determine what is to │
                    │     be worked on     │
                    └──────────┬───────────┘
                               │
                    ┌──────────▼───────────┐
                    │  Ensure all relevant │
                    │ drawings and charts  │
                    │  are available for   │
                    │      reference       │
                    └──────────┬───────────┘
                               │
   ┌────────────────┐   ┌──────▼───────────┐
   │ Identify what  │   │  Check the       │◄──────────────┐
   │ is to be       │   │  approved        │               │
   │ worked on      │   │  voltage         │               │
   └───┬────────────┘   │  indicator is    │               │
       │         YES     │  working         │               │
       │    ┌─────────── │  correctly       │               │
       │    │            └──────┬───────────┘               │
       │    │                   │                           │
       │    │         ┌─────────▼─────────┐                 │
       │    │    YES  │   Satisfactory?   │  NO   ┌─────────┴────────┐
       │    └─────────◄                   ►───────► Repair or replace│
       │              └───────────────────┘       └──────────────────┘
       │
  ┌────▼─────────────┐
  │ Obtain           │
  │ permission to    │
  │ isolate          │
  └────┬─────────────┘
       │
  YES ┌▼──────────────┐ NO
  ┌───◄  Is it in use? ►───┐
  │   └───────────────┘    │
  │                        │
┌─▼──────────────┐  ┌──────▼─────────┐
│ Determine the  │  │ Find out why   │
│ means of       │  │ not            │
│ isolation      │  └──────┬─────────┘
└─┬──────────────┘         │
  │              YES ┌──────▼────────┐ NO
  │              ┌───◄ Has it been    ►───┐
  │              │   │ isolated?      │   │
  │              │   └───────────────┘    │
  │              │             ┌──────────▼──────┐
  │              │             │ Determine the   │
  │              │             │ means of        │
  │              │             │ isolation       │
  │              │             └────────┬────────┘
┌─▼────┐   ┌─────▼──────┐      ┌────────▼──┐
│Isolate│   │Determine why│      │  Isolate  │
└──┬────┘   └─────┬──────┘      └────┬──────┘
   │              │                  │
   └──────────────┼──────────────────┘
                  │
        ┌─────────▼─────────┐
        │ Lock off and post │
        │ notices           │
        └─────────┬─────────┘
                  │
  LIVE   ┌────────▼────────┐  DEAD
  ┌──────◄ Test to confirm  ►──────┐
  │      │ isolation       │       │
  │      └─────────────────┘       │
┌─▼──────────────┐      ┌──────────▼──────────┐
│ Refer to       │      │ Check the approved   │
│ drawing and    │      │ voltage indicator is │
│ charts         │      │ working correctly    │
└─┬──────────────┘      └──────────┬──────────┘
  │                                 │
┌─▼──────────────┐       YES ┌──────▼────────┐ NO
│ Check what has │       ┌───◄ AVI working?   ►───┐
│ been isolated  │       │   └───────────────┘    │
└────────────────┘       │               ┌────────▼────────┐
                         │               │ Repair or replace│
                  ┌──────▼──────┐        └─────────────────┘
                  │ Safe to     │
                  │ work on     │
                  └─────────────┘
```

3

Symptoms and causes of faults

RECAP

Before you start work on this chapter, complete the exercise below to ensure that you remember what you learned earlier.

- In order to carry out fault finding and repair, operatives must be _____ to carry out the work they _____ and ensure the _____ of themselves and _____ .

- Work on or near live conductors may only be undertaken where it is _____ under all circumstances for the conductors to be _____, it is _____ for the work to take place _____ or _____ live conductors and _____ procedures have been taken to _____ injury.

- Safe isolation means making _____ that the supply is _____ off and that it _____ be _____ re-energized.

- Electric shock carries the danger of electrical _____ which occur at the _____ and exit points of the contact and _____ the body along the _____ taken by the current.

- During the fault finding process it is important to _____ the most _____ place to isolate the _____ being worked on.

- In complex installations a _____ to work system may be used to _____ and _____ the work to be carried out.

- Barriers are used to _____ off the work area to create a _____ work environment and prevent _____ personnel _____ the area.

- _____ signs should be placed at the _____ points to the building or location where there is _____ to the public to _____ them of the work on the _____ system.

- To confirm the isolation of a single phase lighting circuit _____ will need to be carried out between all _____ conductors and all _____ conductors and _____ . The _____ Voltage Indicator used to do this must then be _____ to confirm it is still _____ using a _____ unit.

LEARNING OBJECTIVES

On completion of this chapter you should be able to:

- Interpret and apply the logical stages of fault diagnosis and correction work that should be followed

- Identify and describe common symptoms of electrical faults

- State the causes of the following types of fault:

  - High resistance

  - Transient voltages

  - Insulation failure

  - Excess current

  - Short circuit

  - Open circuit.

- Specify the types of faults and their likely locations in:

  - Wiring systems

  - Terminations and connections

  - Equipment/accessories (switches, luminaires, switchgear and control equipment)

  - Instrumentation/metering.

- State the special precautions that should be taken with regard to the following:

  - Lone working

  - Hazardous areas

- Fibre optic cabling

- Electrostatic discharge (friction, induction, separation)

- Electronic devices (damage by overvoltage)

- Information technology (IT) equipment (e.g. shutdown, damage)

- High frequency or capacitive circuits

- Presence of batteries (e.g. lead acid cells, connecting cells).

Part 1 Logical process

This chapter considers the preparation for fault diagnosis and corrective work.

Safe working procedures are important in this process, as we discussed in Chapter 2 of this study book, in the majority of cases the fault finding activity is carried out in installations which are energized and operational. This means there will be a need to consider the work process, our actions and how to ensure the safety of all parties.

The logical stages of fault diagnosis and the subsequent repair work is largely a matter of common sense and a methodical approach.

Identify if there is a fault

It is important that we establish the symptoms of any fault in order to determine whether a fault really exists. If the person reporting the fault is not fully familiar with the normal operation of an item of equipment, it is possible that the perceived fault is actually the normal operation.

Example: 'My fan heater keeps cutting out and if I leave it alone it eventually comes on again for a little while and then cuts off again.' It may well be that the fan heater is controlled by a thermostat and the 'fault' is simply the normal operation of the thermostat.

So first we need to make sure there is actually a fault to be investigated.

Obtain all the information available

Much of the information required for fault diagnosis was discussed in Chapter 1 of this study book. The information that needs to be available will include:

- Any information relating to the events leading up to the fault
- Original drawings and specifications (where available)
- Manufacturers' information (operation and maintenance)
- Maintenance records
- BS 7671 and other relevant standards.

All of this provides background to the installation or equipment and allows us to determine the nature and possible reasons for the fault.

We also need to have a good knowledge and understanding of the equipment, circuits and the protection and control devices involved. Where the fault is on the electrical installation, knowledge of the type of installation and supply system is essential.

In the fault location and repair process, your knowledge and personal experience may be of considerable benefit in achieving a quick and accurate diagnosis and in the swift location of the problem. Experience of the symptoms and common component failure for a particular item of equipment may make the diagnosis and location quicker. This may, however, lead to a fast but incorrect diagnosis and so we should always approach fault finding with an open mind.

For example, in your experience the thermal cut out on a particular space heater frequently fails and the heater stops functioning. When diagnosing the problem with this particular heater it may be an obvious assumption to begin by replacing the thermal cut out, only (flows better) to then find that the heater still does not function and the protective device for the circuit has failed.

The personal experience and expertise of the users of the installation or equipment can be a significant benefit in the diagnosis and repair process. This ranges from the peculiarities of the equipment and common failings, through to the location of controls and the special tools required for access or repair. This information should be obtained, wherever possible, before work begins.

Instrument readings obtained by the client or staff can also provide valuable information. These may be as a result of their initial investigation into the problem or their normal demand figures as recorded on the fixed instrumentation. This may indicate a power surge due to a transient overvoltage, an increase in load or a drop in the supply voltage.

A visual inspection is a good method of establishing the general condition and identifying any obvious problems. For example, a damaged lighting bollard in the car park may be instrumental in causing the loss of power to the car park lights.

Analyze the information

Once the information has been collected we need to evaluate everything we know to determine a suitable course of action. We first have to make sure that the information is complete and accurate and has been correctly recorded so it can be made available as and when required. To decide on the appropriate action we must put to one side, but not discard, any information which appears irrelevant or misleading. During the course of our investigation, our first thoughts may prove to be incorrect and the information will need to be revisited to formulate a different strategy.

Identify the symptoms

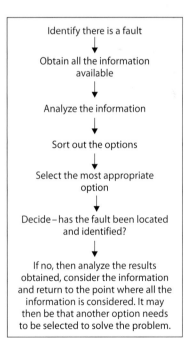

Figure 3.1 *Brief logical process flow chart*

Having determined the first course of action we can begin our investigation to locate the fault. The first check is to confirm that there is a supply available at the circuit or equipment. A basic requirement for the operation of any circuit or item of equipment is that there is a supply and our check would be at the closest and most readily accessible point upstream from the circuit or equipment. This is normally carried out at the item of equipment or the point on the system where the fault has been identified. If there is no supply, the test is repeated at the next convenient point upstream from the original test point and so on.

Example: An item of equipment plugged into a socket outlet circuit is not working. The first check would be to confirm a supply is present at the socket outlet. This could be done using a proprietary socket tester or a test instrument with a suitable adapter.

Megger

Figure 3.2 *Socket tester*

If there is no supply at the socket the next step would be to check another socket on the same circuit, where possible, to confirm if there is no supply to the whole circuit. The check would then be carried out at the distribution board to identify whether the circuit breaker, residual current device (RCD) or fuse has operated and disconnected the supply. Tests would then be carried out to identify why the protective device had operated.

Testing will need to be carried out to determine the type and location of the fault.

Having identified the type and location of the fault the appropriate remedial work is to be carried out to correct the problem. As we discussed earlier this may require some discussion with the client.

Once the remedial action has been completed, testing will need to be carried out to ensure that it is safe for the circuit or equipment to be energized. Providing the test results are satisfactory, the next stage will be a functional test to confirm that everything is working correctly.

Finally, there is the restoration of the building structure and finishes. Depending on the nature of the work involved in the location and the correction of the fault, there may be some remedial work to be carried out to reinstate the building structure or finishes.

This remedial work may be minor but there are occasions when, to access and repair a fault, there is considerable building work involved. It is also possible that the covers on service voids, ceiling tiles and/or grids had to be removed.

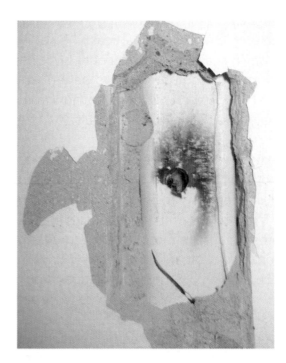

Figure 3.3 *Some structural remedial work will be necessary*

Remember

We are responsible for restoring the building fabric wherever possible on completion of the fault finding and correction.

Task

The motor operated roller shutter doors to a client's loading bay have stopped working and you have been asked to investigate. Draw a simple flow chart to show the actions you would take before beginning the fault finding process.

Part 2 Common symptoms

There are a number of common symptoms for electrical faults which we may encounter. We shall consider those most common symptoms which include:

- Loss of supply
- Reduced voltage
- Operation of overcurrent and fault devices

- Component or equipment failure or malfunction
- Arcing.

Loss of supply

This can be as a result of a failure of the supply to the installation or part of the installation.

The indication of a total loss of supply from the distribution network operator (DNO) or other source is that none of the electrical equipment within the installation operates. This would require some investigation as there are a number of reasons which could have resulted in this condition.

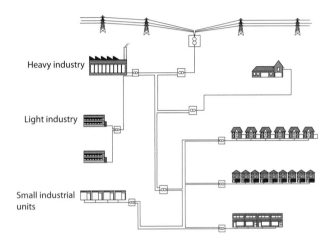

Figure 3.4 *Typical DNO distribution layout*

The DNO has protective devices installed to protect their equipment and cables. The operation of these devices may be as a result of a fault on the network, damage to the distribution network or equipment or the result of an overvoltage on the system. In any event the operation of the DNO protective devices will normally disconnect the supply from a number of premises, although this may not be the case in some rural areas.

This total loss of supply will initially require investigation at the origin of the installation. The DNO equipment is sealed to prevent unauthorized access and so our investigation must take place as close to the origin as possible. This will normally be the incoming terminals of the main isolator.

If there is no supply available at that point, then an enquiry should be made at neighbouring properties to determine whether the situation is local to the one installation or is a general loss of supply. If the supply failure is restricted to the installation we are investigating then it is likely that the fuse in the DNO cut-out has failed.

We would need to:

● Notify the DNO to replace the fuses and
● Carry out some investigation within the installation to try and establish the series of events which caused the operation of the fuses.

The reason for the operation of these fuses will require some careful investigation as the installation may have been overloaded (drawing too much current), or possibly have a fault which resulted in the operation of the DNO's protective device.

Reduced voltage

A reduced voltage fault is where the voltage to the installation, part of the installation or equipment is below that required for the operation of the equipment. The effect will be apparent in a number of ways depending on the voltage present.

The DNO has a maximum variation on the supply voltage given in the Electricity Safety, Quality and Continuity Regulations (ESQCR). This allows a tolerance of +10% to –6 per cent of the nominal supply voltage. On the low voltage public distribution network the declared voltages are 400 V between phases (U) and 230 V between line conductors and earth (U_0).

If the supply voltage falls significantly, the performance of the equipment will be affected and the extent of the problem will depend on the type of equipment. Some equipment is particularly sensitive to voltage drop and motors are fitted with undervoltage protection. This will disconnect the motor from the supply if the voltage falls below a set level and prevent automatic restarting where this could cause danger.

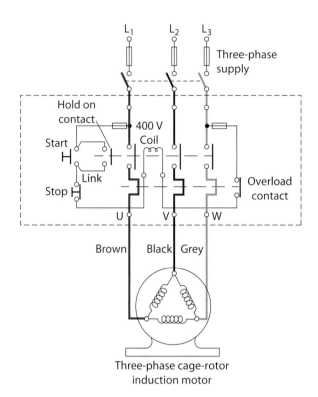

L₁ L₂ L₃

Three-phase supply

Hold on contact

400 V Coil

Start

Link

Stop

Overload contact

U V W

Brown Black Grey

Three-phase cage-rotor induction motor

Figure 3.5 *Coil providing undervoltage protection*

Voltage may fall as a result of the load on the system. This includes not just the load of the installation we are concerned with, but also as a result of the combined load on the network.

The voltage drop within the installation may be caused by the resistance of the conductors within the installation. During the design process the designer will select conductor sizes to ensure that the voltage drop stays within acceptable limits.

BS 7671 gives the maximum voltage drop within the installation. This is measured from the origin to the furthest point electrically, that is not necessarily physically on each circuit.

Voltage drop		
Low voltage installations	Lighting	Other circuits
Supplied from a public DNO	3%	5%
Supplied from a private source	6%	8%

Figure 3.6 *Maximum voltage drop for low voltage installations taken from BS 7671*

However, changes to the installation or equipment may result in an increase in load which in turn will create an increase in voltage drop. On complex installations, changes to equipment can have a significant effect without necessarily causing an overload of any circuit.

Poor performance and/or overheating of equipment are common signs of excess voltage drop.

Note

More detailed information on selection of conductors for voltage drop can be found in the *Planning and Selection for Electrical Systems* study book in this series.

Operation of overcurrent or fault current devices

A common fault is the loss of supply to equipment, circuits or installations as the result of the operation of overcurrent or fault current devices. These devices include fuses, circuit breakers and RCDs.

The term overcurrent includes both overload and short circuit currents. Fuses and circuit breakers are used to provide protection against overload and short circuit currents.

Remember

An overload occurs in a healthy circuit and current often builds up over time; whereas a short circuit occurs quickly and causes a high current to flow.

Where additional protection is not required and the earth fault loop impedance of the circuit is suitably low, fuses and circuit breakers may also be used to provide fault current protection.

Note

More detailed information on the selection and use of protective devices can be found in the *Planning and Selection for Electrical Systems* study book in this series.

The operation of fuses and circuit breakers will leave parts of the installation without supply and the location within the installation will determine how much of the system is without supply. If we consider the distribution layout in Figure 3.7 we can see that the operation of a protective device at:

● Switchfuse 2: will cause the item of equipment B to stop functioning
● CCU 1: will cause all three items A, B and C to stop functioning
● DB2: will cause all the circuits supplied from CCU 1 to stop functioning
● DB1: will cause all the circuits supplied from DB2 to stop functioning.

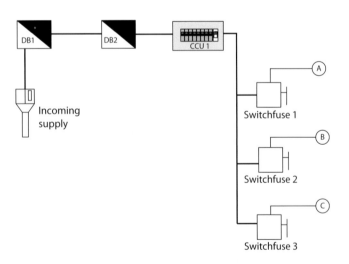

Figure 3.7 *Distribution block diagram*

The operation of a protective device may be caused by an overload, a short circuit or, in many cases, an earth fault. Having identified the reason for no supply we must then establish the cause.

RCDs and residual current circuit breakers with over-current protection (RCBOs) (residual current devices with overload protection) are used to provide protection against fault current to earth. These devices monitor the current in the live conductors and if a predetermined imbalance is detected, the device automatically disconnects the supply. However, the simple fact that the device has operated does not necessarily mean an earth fault exists as any imbalance will cause the operation of the device.

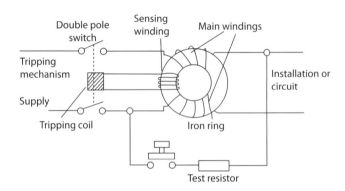

Figure 3.8 *Double pole RCD circuit*

It is not uncommon for inductive loads to cause the operation of an RCD during start up or shut-down. Once again having determined the reason we will have to identify the cause of the device operation.

Note

More detailed information on RCDs can be found in the *Planning and Selection for Electrical Systems* study book in this series.

Equipment failure or malfunction

If we have determined that there is a supply to a circuit or item of equipment and that the voltage present is within the manufacturer's tolerance for the equipment, it follows that there must be a failure of either the equipment or a component within it. This will then lead us to further investigation and testing within it to identify the cause of the failure or malfunction.

For example, if an item of equipment runs in the forward drive but fails to operate in reverse there are a number of possible options to consider. It is apparent that the fault lies with the circuitry for the reverse operation so the logical following of the circuit and testing for voltage and measurements, in accordance with the manufacturer's information, will enable the fault to be located.

Arcing

The normal operation of switches and contacts in electrical circuits produce arcing at the terminals as the contacts open. The presence of arcing is generally identified using the senses of hearing and smell. As time goes on, and the more use the contact gets, the clearance distance between the contacts changes due to the arc deposit on the surface of the contacts. The contact resistance also becomes higher as the carbonized deposit builds up.

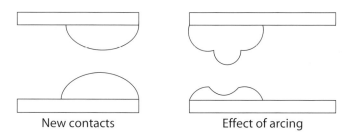

| New contacts | Effect of arcing |

Figure 3.9 *The effect of arcing on contacts*

The use of solid state controls such as the soft-start for inductive motors has reduced the heavy switching loads experienced by older machines. However, the effects of arcing can be detected on a light switch controlling discharge lights as they are switching an inductive load. The crackle as the lights are switched is generally much

more noticeable when switching off the load, as the contacts are breaking the load current and the characteristics of an inductive load makes this more onerous.

Note

More detailed information on inductive circuits can be found in the *Principles of Design, Installation and Maintenance* study book in this series.

Poor terminations present a considerable risk and are another cause of arcing. Loose terminations, as we discussed earlier, can create a real risk of fire.

Figure 3.10 *One result of a poor termination*

Where arcing takes place there is a distinctive metallic smell produced and this can often be detected without the removal of covers.

The number and variety of symptoms and faults which may occur in an electrical installation are many and it is not possible to cover all the possibilities in a work book such as this. The process of fault finding does require experience of the installation type or system and the faults which may occur.

We will consider some of the possible faults related to connected accessories and equipment.

Equipment or accessory	Symptom	Possible cause
Switched socket outlet	Equipment not working	Faulty switch
Two-way switch	Incorrect operation not always able to operate lights from other switch	Strapper and common connections cross connected
BC lamp holder	Lamp only works when held upright	Damage to supporting clips or sticking pins
40 W fluorescent lamp	Flashes but does not light	Failed starter
Sodium discharge lamp	Glows red but does not light	Failed igniter
One-way switch	Crackles when operated	Worn switch contacts/weakened spring
Electric radiator	Stopped working	Thermal cut out has operated
Lighting circuit	As more lights are switched on lights dim	Lights incorrectly connected in series
Electric cooker	Oven works but fails to reach temperature	One failed element
Fridge	Light works but fridge does not operate	Faulty thermostat
Consumer unit	Metallic smell	Poor termination arcing
Washing machine	Does not empty water	Pump motor failed

Figure 3.11 *Typical fault symptoms and possible causes*

Task

Identify a possible fault for each of the items in Figure 3.12.

Equipment or accessory	Symptom	Possible cause
Socket outlet	Works intermittently	
Dimmer switch	Light does not operate	
ES lamp holder	Lamp flickers	
Lead-lag fluorescent light	One tube fails to light	
Electric oven (auto cook)	Does not heat up	

Figure 3.12

Part 3 Common causes

There are common fault conditions which we are likely to encounter in electrical installations. These include:

- High resistance
- Transient voltages
- Insulation failure
- Excess current
- Short circuit
- Open circuit.

Whilst there may be a number of reasons for these faults to occur we will consider some of the most common.

High resistance

These faults are generally taken as occurring over the range of continuity values. The continuity of conductors is measured in mΩ/m and typically a 2.5 mm^2 copper conductor will have a resistance of 7.41 mΩ/m (0.00741 Ω). With these values of resistance a high resistance could easily be in ohms.

The most common reasons for high resistance faults are poor terminations and the failure of components. Terminations may be made which secure the conductor within the terminal but fail to make a sound electrical connection.

Over time this is likely to lead to overheating and arcing but the immediate effect is likely to be an increase in the conductor resistance for the circuit.

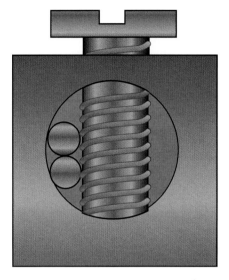

Figure 3.13 *Conductor held but not electrically sound*

A similar effect is caused by the failure to tighten the termination sufficiently and this will produce similar results. Oxidization of contacts in switches and contactors can also introduce high resistances.

As components in equipment fail, their resistance can increase, in much the same way as switch contact resistance can rise over time. Electronic devices may exhibit these characteristics when they begin to fail.

Transient voltages

Transient voltages are very short periods of high overvoltage which are generally caused by switching on the system or induced from the atmosphere. The latter is normally due to lightning and static charges which are built up in the atmosphere. BS 7671 identifies the requirements for protection against these transient overvoltages and the installation should be suitably protected. This is often achieved through the construction of the installed equipment.

Voltages generated in the atmosphere may induce transient voltages on the supply network and these then appear at the installation. There does not need to be a direct lightning strike on the supply system; a ground strike in the vicinity can produce transient voltages on the supply.

Transient voltages can also be produced as a result of switching on the network or within the installation. An inductive load produces a back electromotive force (emf) as part of the normal operation. When an inductive load is switched off, this back emf tries to continue the flow of current and can produce very high voltages as the back emf collapses.

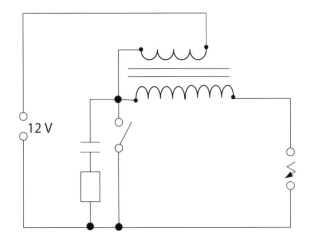

Figure 3.14 *Simple ignition coil circuit*

This is the principle used in the ignition coil for petrol engines, where the coil is supplied at 12 V and when the circuit to the coil is opened, a voltage up to 40 000 V is produced as a result of the collapsing emf.

Switching inductive loads in electrical installations can produce a similar effect and the larger the inductive load the greater the potential transient voltage.

These transient voltages can seriously damage equipment which is not suitably protected and many installations now include surge protection devices to protect the installation and equipment.

Insulation failure

Insulation failure is generally caused by either damage or degradation. Most electrical insulation used in cables and conductors is susceptible to mechanical damage and that is why additional mechanical protection in the form of a sheath, armour or a separate enclosure is essential. However, conductor insulation can be damaged at termination points and, where cores are exposed within accessories and equipment.

Insulation does not have to be cut through to be damaged; pressure on the insulation is often enough to produce a breakdown. For most insulation the thickness of it is essential to the level of insulation provided. Where the insulation is crushed and the thickness subsequently reduced, the level of insulation falls.

Figure 3.15 *Insulation damaged by crushing*

Older cables, particularly those with rubber-based insulation such as VRI and TRS (vulcanized rubber insulated and tough rubber sheath) deteriorate over time. The time it takes for the degradation depends on a number of factors such as load and environment. The insulation generally becomes brittle although it may retain its insulating properties until it is disturbed.

The properties of most insulation materials do deteriorate naturally over time although the thermoplastic and thermosetting insulation materials tend to have a longer life span than the rubber based insulation.

Disturbing or moving the cable often results in the insulation cracking or coming away from the conductors completely.

Figure 3.16 *Rubber insulation breakdown at a termination; pieces of insulation can be seen in the switch box and the conductors are bare*

During the normal operation of the installation, factors like vibration, structural movement and environmental changes can all cause the cables to move and flex, potentially resulting in a breakdown of insulation. The natural

degeneration of the insulation material will also result in a falling insulation resistance over time.

Excess current

Excess current may be as a result of an overload, short circuit or a fault to earth. The overload is the only one which does not occur as the result of a fault on the installation or equipment. It is commonly the result of increasing the load on a circuit or installation. This may be caused by installing additional equipment or by changing existing equipment for a higher rated replacement.

Some items of equipment fail towards closed circuit, resulting in an increased current flow. For example, some tungsten filament lamps often draw a high current as the filament fails causing the protective device to disconnect the supply.

Short circuit

A short circuit is due to a fault of negligible impedance occurring between live conductors. This can be the result of failing insulation or components in the circuit. The failure of insulation is a common cause of a short circuit and in older installations is often due to the degradation of the insulation over time. Mechanical damage to the insulation can also result in a short circuit between live conductors. The failure of some components can also lead to a short circuit.

Open circuit

This may be the result of a break in a conductor although this will require some form of mechanical damage or termination failure. The open circuit fault due to such failures often manifests itself with sections of circuits or equipment that are no longer working. It is likely that an open circuit will result from the failure of a component or accessory. A common example is a fuse or thermal cut-out which disconnects in the event of overcurrent or overtemperature respectively.

The open circuit fault generally does not result in the operation of a protective device; however, the effects may be similar in that the circuit does not operate.

Task

For the events and symptoms identified in Figure 3.17 identify one possible cause of the fault.

Event	Symptom	Possible cause
New wall cupboards fitted in a domestic kitchen	Socket outlet circuit breaker operates immediately it is switched on	
Additional 500 W flood-light added to outside circuit in car park	Circuit breaker operates after lighting has been on for a short period.	
Socket outlet circuit in a commercial kitchen	RCD operates at irregular periods	
New suspended ceiling installed in a small office	Only the first six lights operate on one circuit.	

Figure 3.17

Part 4 Common faults and locations

Having discussed a number of types of faults we need to consider the likely faults within the electrical installation and equipment where these are likely to occur. We will begin with the possible faults in wiring systems.

Wiring systems

Faults in the wiring system relate to the cables and conductors forming the system and may be influenced by their connection with components and equipment. Wiring systems vary between those that are relatively simple and those which are extremely complex. However, the type and location of the faults are generally similar.

The most likely faults to occur in conductors and cables are short circuit and open circuit faults. These generally occur as a result of deterioration or damage. In the case of the deterioration of the cable or conductor insulation, the fault is most common at points of termination where the sheath of the cable is removed and the insulation exposed. Any heat which is generated by the equipment or accessories will often accelerate the deterioration of the insulation. Changes to accessories or equipment may cause the insulation to fail at, or near the termination.

Damage to insulation often occurs where the cables enter and are terminated, so some of the locations for both damage and deterioration are the same. Similarly, damage may occur from physical abuse and depending on the location, damage by fauna and flora.

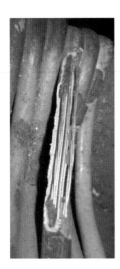

Figure 3.18 *Rodent damage to cables*

Terminations and connections

Terminations and connections in cables and conductors are an inherent weak spot unless the appropriate type of connection is used. Poor terminations can result in poor performance, arcing and fire. In some locations the terminations are subjected to external influences which result in even sound terminations being at risk. The connections to machinery and movable equipment are prone to vibration and may become loose, particularly where screw terminals are used.

In general, terminations which are not subject to such external influences and are properly constructed will remain trouble free.

Figure 3.19 *Result of poor termination*

Equipment and accessories

Equipment and accessories used in adverse conditions such as wet and corrosive locations may deteriorate as a result of their everyday use. Areas such as car washes will require particular care with the terminations as they are liable to vibration, water and cleansing chemicals.

Switchgear and control equipment may also become faulty during operation. The poor quality termination problems exist with all equipment but there are additional possible faults. Mechanical switches which are used to control energized circuits are making and breaking under load conditions. The making, and more significantly the breaking of load current produces heat and arcing at the contacts.

Over time functional switches may suffer the damage at the contacts mentioned earlier. This can result in the switch having a high resistance connection and the subsequent problems this will cause. Switches also rely on the mechanical components for their operation and these can also fail. Many switching functions from large fused switches to the humble pull switch, rely on springs to make and break contact quickly. Over time these springs may weaken and fail to function correctly when making and breaking the contacts. This increases the wear on the contacts and accelerates it.

Electronic switching devices such as dimmer switches are rated to control a maximum specific load. If this is exceeded then the heat generated during the operation of the circuit increases and the electronics suffer as a result. This may result in, for example, the dimming function failing, whilst the on/off operation of the switch still works mechanically.

Often a simple item of equipment such as a tungsten filament lamp can produce faults on a circuit. Failure of the filament often results in a high current being drawn as the element parts drawing an arc. This in turn causes the protective device to disconnect whilst there is no damage to the installation or equipment.

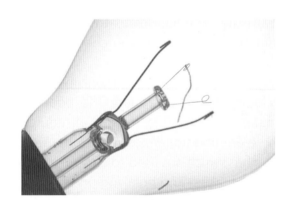

Figure 3.20 *Failed tungsten lamp*

Instrumentation and metering

Installations supplied by the DNO will be metered to record the energy consumption. In addition to this metering there are many other points within installations which may be metered or monitored

using instruments. Measurement of voltage, current and power is quite common in complex installations for particular items of equipment. Many instruments are installed via current or voltage transformers so they do not form part of the installation supplying the equipment. This means that in the event of a failure the supply to the installation or equipment is not interrupted.

For directly connected instruments the situation is different.

Voltage is measured in parallel to the circuit and so in the event of the voltmeter going open circuit, the equipment will continue to operate. In the case of a short circuit however, the protective device would operate and disconnect the supply.

Current would be measured in series with the equipment and so the situation would be the reverse for an ammeter. An in-line ammeter which goes open circuit will disconnect the supply from the equipment. A short circuit in the same instrument will not interrupt the supply to the equipment.

In many cases the current to be measured is higher than is practical for the use of an in-line ammeter to be used. Therefore current transformers are commonly used to measure the current.

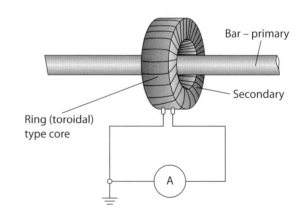

Figure 3.21 *Typical current transformer used for instrumentation*

This effectively allows current to be measured using a bar primary transformer and secondary winding. As a result, a large current may be measured by a small instrument with the scale suitably adjusted.

Remember

If an ammeter which is supplied via a current transformer goes open circuit a high voltage will be present across the instrument terminals.

Note

More detailed information on transformers and their use can be found in the *Principles of Design, Installation and Maintenance* study book in this series.

Before we finish this section we will look at some more of the faults which may occur on items of equipment and the likely causes.

Equipment	Symptom	Possible cause
Storage water heater	Heats up and boils the water	Thermostat not operating leaving element connected all the time
Single phase ac motor	Fails to start and makes a humming noise	Centrifugal switch open circuit
Three phase motor	Starts and runs but gets very hot	One phase of the supply has failed
RCD protecting a ring final circuit	The RCD trips when there are no loads connected	Possible that the RCD is faulty
Fluorescent luminaire	Tube is blackened at end and fails to light	The electrode filaments in the tube ends burnt out

Task

List two common faults which may occur on each of the following:

a the cables of a lighting circuit

_____

_____

_____

b a socket outlet circuit protected by an RCD

_____

_____

_____

c a two-way light switch

_____

_____

_____

d a 3 kW heater.

_____

_____

_____

Part 5 Special precautions

There are a number of occasions during fault finding when special precautions need to be taken regarding the work activity, the location or the material and equipment used.

Lone working

It is common practice in the electrical installation industry for electricians to work alone on many occasions. The statutory regulations do not require persons to be accompanied and so a risk assessment must be made. This will determine whether the presence of a second person will improve the safety of those involved.

There are some obvious occasions where the presence of a second person would improve the safety of the electrician including where work is carried out:

- On or near live conductors
- In areas occupied by other people or the general public
- In confined or restricted access areas
- In hazardous areas (other than the electrical hazards).

Working on or near live conductors should always be avoided wherever possible and in almost all cases in electrical installations it is unnecessary. However, accidents occur when live working has to be undertaken, often due to distractions or interference from other people. A second person could minimize these risks.

Where testing is to be carried out and access to live terminals is necessary, there could also be a danger to other people in the vicinity. Barriers and warning signs will be necessary to warn and keep them away from the area. However, these warnings may be ignored and a second person would be able to prevent any outside interference.

When working alone it is important that we carry out a risk assessment and take the necessary precautions to prevent danger to ourselves and other people. This will include the use of barriers and warning signs to advise people of the work activity and keep them away from danger. These may be needed where the work involves:

● Having open access to holes in floors, service ducts, etc.

● Using steps or ladders

● Activities in thoroughfares such as corridors, access routes and stairwells

● Working on or near live conductors or equipment.

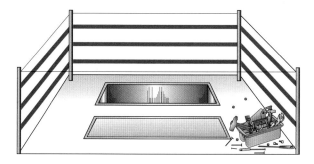

Figure 3.22 *Place barriers around open floor access points*

It is important to remember that whilst you are working there is no one else to look out for possible dangers from or to yourself, or other people.

Remember
A second person does not have to be skilled in the work being carried out. They do need to be aware of the possible dangers involved and how to prevent subsequent injury.

A further important action is to make sure someone, preferably a supervisor or responsible person, on the premises is aware of:

● Where you are working

● What is involved

● How long the work is likely to take you.

It is also important that you inform this person when the work is completed or if you change your location. Having someone aware of the activities allows a check to be made to confirm all is well at any time during the work. Should you be injured someone knows where you are and can check on you.

Hazardous areas

The hazards may be from many sources and the precautions are varied depending on the circumstances. The statutory regulations require that hazards are removed wherever possible. The 'hazardous area' will already have all these considerations taken into account but do not necessarily reduce the risk.

Some of the hazardous areas are due to the materials used and some to the activities carried out. These areas include:

● Petrol stations

● Fuel and gas depots and processing plants

● Flour mills and bakeries

● Wood machine shops

● Food processing plants

● Paint spray booths

● Laboratories

● Manufacturing premises.

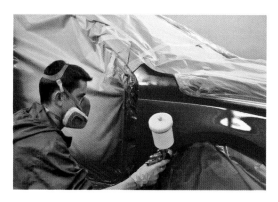

Figure 3.23 *Spray booths are hazardous areas*

There will be specific requirements for working in hazardous areas and these will depend on the location and activities.

It is common to have a permit to work systems in operation as they need to be controlled responsibly to ensure the safety of everyone involved and the operation of the premises. In addition there may be special requirements; for example, in explosive atmospheres intrinsically safe equipment will be required if testing is to be carried out in the area.

In some manufacturing and pharmaceutical preparation areas it may be necessary for you to be accompanied at all times to prevent entrance to restricted or dangerous areas without suitable protection. In some laboratories radiation may be an issue and exposure recorders will need to be worn.

The common rule is to establish what the hazards are, what measures are implemented for safety and make sure you comply with them.

Fibre optic cabling

Carrying out work on fibre optic cables introduces a number of hazards and when fault finding, these cables are often in service, creating additional problems. There are some basic rules which, if followed, should protect you from the risks involved.

Basic rules when working with fibre optic cables:

1   Protect your eyes from light: Never look at the end of a fibre to see if it is working.

Whilst the light levels are relatively low they are concentrated and exposure can lead to permanent damage.

2   Use a fibre optic power meter to confirm whether a fibre is in use.

3   Wear eye protection when terminating fibre optic cables and these should have side shields to prevent particles of the glass entering the eyes.

4   Do not touch your eyes or eat or drink whilst working with fibres.

5   Wash your hands after handling fibres.

6   Account for all bare fibres: these need to be placed in a sharps container, similar to those used for hypodermic needles.

7   Work on a surface which is dark in colour to provide contrast to the fibres so these can be easily seen.

8   When working with epoxy keep the fibre and the area clean using an alcohol wipe to clean the fibre.

9   Wash your hands after working with the epoxy to remove any traces. Do not use an alcohol wipe on your hands as this will spread the epoxy on the skin and not remove it.

10   Keep the work area clean and dust free.

The offcuts of fibre can easily penetrate the skin and are difficult to see and remove. They break off readily even when extracted with tweezers and can cause infection. Similarly, they can enter the eye and cause serious problems and may result in permanent loss of sight. If the fibres are ingested through eating, drinking or smoking they can cause internal bleeding.

Providing the manufacturer's instruction and the simple safety rules are followed these cables can be safely terminated.

a)

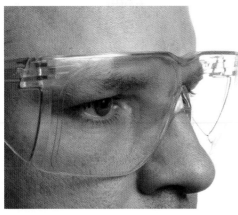

b)

Figure 3.24 *Disposal bin and wraparound eye glasses*

> **Note**
>
> More detailed information on the termination of fibre optic cables can be found in the *Termination and Connection of Conductors* study book in this series.

Electrostatic charge

Electrostatic charge is generally referred to as simply 'static' and these electrostatic charges are built up in a number of ways. The methods of producing these charges are friction, induction and separation. Friction is the one most commonly experienced by people and is caused by rubbing materials together. Walking across carpeted or plastic tiled floors will generate an electrostatic charge in people. When they then touch a material which is at a lower potential than their own electrostatic voltage a discharge of the stored energy occurs. This is known as electrostatic discharge (ESD). This effect takes place between charged materials and earth, which is generally considered to be at 0 volts.

Walking through the house on the carpets and receiving an electric shock when we go to touch the door handle or switch on the light, are common examples of this phenomenon.

The material that accepts electrons becomes negatively charged and the material that gives up electrons is left positively charged.

Separation of materials can also produce an electrostatic charge and this is evident when pulling cling wrap from a roll. The act of separating the layers of film generates an electrostatic charge and the most common result is that the film wraps itself onto the person's hand and needs to be peeled off.

Induction is best explained by way of the example, i.e. using electrostatic charge to stick a balloon to a wall. The balloon is normally charged by friction, such as against a woollen cloth. When it is placed against the wall the material of the wall surface is polarized by the charged balloon to produce an equal and opposite potential, thus the two surfaces are attracted and the balloon sticks to the wall. The charge on the wall is produced by induction.

Remember

The human perception level of ESD is 3000 V (3 kV) and values below this level are regularly produced but the individual is not aware of the charge or the resulting ESD.

Electrostatic charge is built up in insulators which can generate, store and discharge energy. Whilst ESD may give a shock and cause a reflex action it is unlikely in normal circumstances to cause a fatal electric shock. It may cause injury as a result of the reflex action which creates dangers like falling from steps or ladders.

Figure 3.25 *Electrostatic charges can be amusing*

A spark from a charged surface to an operative may not normally be particularly dangerous. However, in an explosive or flammable atmosphere the result of a relatively insignificant spark could be disastrous.

Sensitive electronic equipment and components are highly susceptible to ESD and at voltages which will not be perceived by persons.

Metal conveyor belts, tensioners and idler rollers can become charged when plastic gears isolate the conveyor belts from earth and greased bearings may isolate the metal drums and rollers.

When carrying out fault finding, the most significant areas requiring particular precautions are the effects of ESD on us and equipment which is sensitive to ESD.

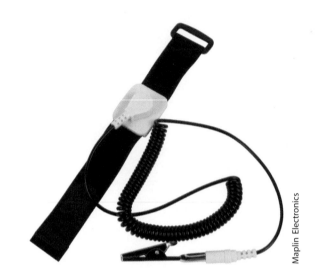

Maplin Electronics

Figure 3.26 *Anti-electrostatic earthing strap*

The protection of sensitive equipment is generally achieved by the use of an earth strap which is worn during the work activity. This connects the operative to earth whilst they are working, and so discharges any electrostatic charge before it can build up. Antistatic mats are also used when placing components on a work surface to ensure that they are not affected by any residual charge.

Electronic devices

Electronic devices and equipment may easily be damaged by overvoltages. When fault finding within an electrical installation it is common to carry out tests of insulation resistance. These tests are carried out at voltages in excess of the normal operating voltage; the actual voltage is dependent upon the circuit being tested.

Nominal circuit voltage	Test voltage (dc)	Minimum insulation resistance
SELV & PELV only	250 V	0.5 MΩ
Up to and including 500 V (excluding SELV and PELV only)	500 V	1.0 MΩ
Over 500 V	1000 V	1.0 MΩ

Figure 3.27 *Insulation resistance test applied voltages and minimum acceptable values*

When such tests are carried out, any electronic equipment may be damaged by the applied test voltage. It is important to determine whether any such equipment is involved before carrying out testing which may introduce higher than normal operating voltages.

If there is any doubt it is advisable to test at a lower voltage initially to confirm there are no items of equipment still connected, as loads or equipment will give very low readings. However, when fault finding these very low readings may be due to a fault so it is not easy to identify components in this way.

Devices which may be affected include:

● Capacitors
● Smoke detectors
● Dimmer switches
● Electronic controllers
● Passive infrared sensors
● Control equipment
● IT equipment and computers
● Photocells.

Any electronic equipment or equipment containing electronic components should be identified, isolated and where necessary, linked out to prevent damage from excess voltages.

IT equipment

IT equipment includes telephone systems, printers, plotters, fax machines, computers and data terminals. As mentioned in the electronic equipment section above, all this equipment may be damaged by excess voltage, particularly during testing.

In addition there is also the need to consider shutting down this equipment when fault finding. Before any IT equipment is turned off, permission must be obtained to ensure that no loss of information occurs. Permission to shut down should come from the person responsible for the installation, not just any operative or the one working on the equipment you need to isolate.

Figure 3.28 *Typical IT equipment*

Shutting down IT equipment must be done correctly to ensure information is saved and that other activities, information or data storage is not affected. It is possible to cause considerable data loss and associated costs for the customer if the shutting down is not done correctly. In some cases the costs associated with incorrect shutdown procedures can run into millions of pounds. Not only can the data be lost but also customers and thousands of hours of work so severe commercial disadvantage may result. Organizations that trade in data can lose millions of pounds a minute in the global market.

Remember

Always seek permission before shutting down IT equipment. Ask the responsible person to have the system, or part of the system, shut down. Do not simply isolate the circuit.

High frequency circuits

In electrical installations, high frequency circuits are generally found in discharge light fittings as high frequency improves efficacy (lumens per watt) and removes the stroboscopic effects of this lighting. For tubular fluorescent tubes the frequency is generally around 30 kHz but must be above 16 kHz; below this level they are audible to the human ear and a high pitch whistling may be heard.

The higher frequency is created electronically and this also involves some filtering circuitry. The implication for the circuit supplying the luminaires is that there may be leakage currents flowing in the cpc of the circuit under normal operation.

This means that the cpc is carrying current during normal operation and should the cpc be disconnected whilst the circuit is operating, the exposed conductive parts will be above earth potential and pose a risk of electric shock.

Capacitive circuits

When carrying out insulation resistance, testing the conductors are tested using a dc voltage and generally at 500 V. Parallel conductors insulated from one another which have a dc voltage applied is the standard construction for a capacitor. When this test is carried out, one conductor becomes positively charged and the other negatively. Once the test is completed the conductors can hold this charge for some time and there is a risk of electric shock and arcing as a result. The conductors should be discharged on completion of the test to remove this danger.

Capacitors are used in many circuits for power factor correction and starting and these are fitted internally with a discharge resistor. The purpose of this resistor is to allow the capacitor to discharge when the circuit is isolated and so remove the risk of electric shock from the capacitor discharge. Should this resistor become open circuit, the capacitor will retain the charge but there will be no outward signs that there is anything wrong. It is advisable to short circuit the capacitor terminals once the supply has been safely isolated to remove any potential shock risk.

Figure 3.29 *Typical fluorescent fitting power factor correction capacitor*

Large banks of capacitors may be used to provide power factor correction for an installation, with the actual connected capacitance controlled by a monitoring and control system. Considerable care needs to be taken when working on installations with this type of system in place.

In some electronic circuits, capacitors are used to provide a short term continuous supply for timers and the like. These do not have discharge resistors as they are intended to discharge through the timer to maintain the timer operation, and keep time settings should the supply fail. Whilst these are generally small capacity supplying very small loads, the risk of electric shock remains. Where there is a visible time reference and this continues to be operational when the equipment is isolated, there is likely to be a capacitor backup.

> **Remember**
>
> Capacitors store energy in a similar way to batteries and they should be treated with care.

> **Note**
>
> More detailed information on capacitors and their use can be found in the *Principles of Design Installation and Maintenance* study book in this series.

Presence of batteries

Batteries provide a source of storable energy and are used in many locations where either a break in supply could result in potential loss of data (uninterruptable power supplies for example), or where equipment needs to continue to work in the event of a power failure (emergency lighting for example). It is important to remember that the battery is a source of energy and so when connecting or disconnecting batteries

there is a risk of electric shock. Whilst the voltage may be low for individual batteries, when connected in series this voltage will rise.

> **Remember**
>
> Batteries connected in series increases voltage. Batteries connected in parallel increases the current available.

There is a lot of stored energy in a battery and so short circuiting the terminals will result in a sudden discharge of the stored energy. This can result in a considerable arc which may result in serious electrical burns. A spanner accidently dropped across the terminals of say, a car battery, can produce a considerable arc with a lot of heat and molten metal.

Figure 3.30 *Typical sealed-for-life battery*

Most modern batteries are sealed-for-life gel-type batteries which require no routine maintenance. But many batteries in service are lead-acid-type batteries which need to be regularly maintained.

These batteries give rise to a number of dangers, the first being that during their operation they give off hydrogen gas which goes into the atmosphere. The rooms in which these batteries are stored must be well ventilated to prevent a build-up of this explosive gas. There should be no smoking or naked flames in the location and care must be taken not to create arcs during connection and disconnection of individual batteries.

In addition to this the batteries need to be checked to make sure the electrolyte is to the correct level and the specific gravity is correct. This may require the batteries to be topped up so sulphuric acid and distilled water are used to form the electrolyte. This may be provided pre-mixed or mixed on site. Always add acid to the distilled water never the other way around.

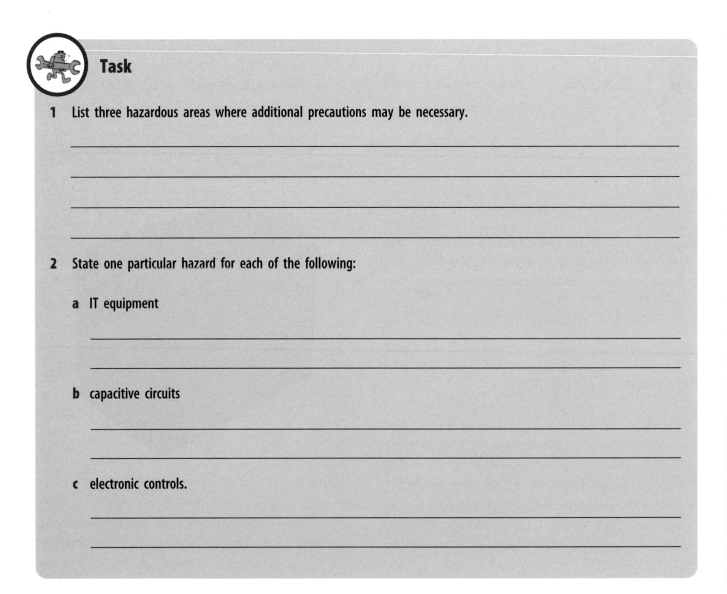

Task

1   List three hazardous areas where additional precautions may be necessary.

_____

_____

_____

_____

2   State one particular hazard for each of the following:

a   IT equipment

_____

_____

b   capacitive circuits

_____

_____

c   electronic controls.

_____

_____

Congratulations, you have now finished this chapter, complete the self assessment questions before going onto the next chapter.

SELF ASSESSMENT

Circle the correct answers.

1   A fault has been reported on an electric oven which is not working. On arrival at the site the first action to be taken will be to:
   a.   isolate the supply
   b.   confirm the fault exists
   c.   replace the oven thermostat
   d.   fit new heating elements to the oven

2   A chop saw supplied from a 13 A socket outlet has stopped working. Other equipment supplied on the same circuit is still operational. The likely cause of the chop saw not working is:
   a.   a short circuit within the socket supplying the saw
   b.   the supply to the installation has been disconnected
   c.   the circuit protective device has operated
   d.   the fuse in the 13 A plug has operated

3   The damage shown in Figure 1 is likely to have been caused by:
   a.   an overload
   b.   a short circuit
   c.   poor termination
   d.   adverse environment

Figure 1

4   New fitted wardrobes have been installed in a hotel bedroom. When the room is checked for the guests, the socket outlets are not working and the circuit breaker cannot be reset. The most likely fault is the:
   a.   circuit breaker is faulty
   b.   socket outlets are faulty
   c.   circuit is open circuit
   d.   circuit cable has been damaged

5   The equipment shown in Figure 2 is used to prevent:
   a.   earthing sensitive equipment
   b.   electrostatic discharge
   c.   electric shock
   d.   earth faults

Maplin Electronics

Figure 2

Progress check

Tick the correct answer

1. **A fault on a power circuit has been located and a damaged length of cable replaced. The document which should be issued to the client following this repair is a:**

   ☐ a. Minor Electrical Installation Works Certificate

   ☐ b. Electrical Installation Condition Report

   ☐ c. Maintenance schedule

   ☐ d. Manufacturer's warranty

2. **During the initial discussion with the client regarding the work involved in locating a fault, one of the key areas to be considered is the need for:**

   ☐ a. Replacement parts

   ☐ b. Access arrangements

   ☐ c. Total cost of the work

   ☐ d. Materials required

3. **Communication with the client regarding fault diagnosis should always be clear, accurate and:**

   ☐ a. Courteous

   ☐ b. Written

   ☐ c. Verbal

   ☐ d. Technical

4. **A fault has been reported on a vacuum forming machine. The information that would be required to carry out the actual fault finding includes the:**

   ☐ a. Manufacturer's warranty

   ☐ b. Electrical Installation Certificate

   ☐ c. Maintenance records

   ☐ d. Manufacturer's instructions

5. **All the necessary information and documents for a faulty item of equipment and the electrical installation supplying it are available. The most likely effect of this on the fault finding process is that it will:**

   ☐ a. Cost more

   ☐ b. Take less time

   ☐ c. Require replacement parts

   ☐ d. Have to be done out of hours

6. **The information which may be obtained from the client to establish the condition of a pillar drill prior to it becoming faulty is the:**

   ☐ a. Minor Electrical Installation Works Certificate

   ☐ b. Electrical Installation Certificate

   ☐ c. Maintenance schedule

   ☐ d. Manufacturer's warranty

7. The repair of a laser cutter requires a bespoke part which is not held in stock but has been ordered. The information the client requires is the detail of the:

☐ a. Supplier of the parts

☐ b. Expected delivery date

☐ c. Difficulty in obtaining the parts

☐ d. Part number and reference

8. A current flowing across the chest of a person which is likely to result in ventricular fibrillation is generally of a magnitude of:

☐ a. 20 mA

☐ b. 30 mA

☐ c. 40 mA

☐ d. 50 mA

9. When confirming safe isolation of a single phase motor the total number of tests to be made is:

☐ a. 3

☐ b. 4

☐ c. 7

☐ d. 10

10. A three phase machine is to be isolated from a local isolator to enable fault finding to be carried out. This process includes switching off the isolator, confirming the motor supply is dead and:

☐ a. Notifying the client

☐ b. Putting barriers in place

☐ c. Placing warning notices

☐ d. Locking off the circuit breaker

11. Once isolation has been confirmed, and before fault finding begins, the test equipment must be confirmed to be:

☐ a. Compliant with GS 38

☐ b. Working, using a proving unit

☐ c. Suitable for the expected voltages

☐ d. In calibration

12. The maximum length of exposed metal tip for test probes, used to confirm safe isolation given in GS 38 is:

☐ a. 2 mm

☐ b. 4 mm

☐ c. 6 mm

☐ d. 8 mm

13. A conveyor system has been isolated to enable fault finding to be carried out. Whilst the system is isolated an engineer is carrying out some mechanical maintenance. The correct security arrangements for the isolation are that each operative has:

☐ a. A key for a common padlock

☐ b. An individual padlock with a common key

☐ c. Their own unique padlock and key separate from all others

☐ d. A key for a common padlock and personal warning notices

14. Which of the following is not an item of information required when establishing the presence of a fault:

☐ a. Electrical Installation Certificate

☐ b. Events leading up to the fault

☐ c. Manufacturer's information

☐ d. Maintenance records

15. An overload fault occurs in a circuit:

- ☐ a. Instantly
- ☐ b. Over a period of time
- ☐ c. As insulation deteriorates
- ☐ d. When equipment is replaced

16. A fault on an air compressor causes the RCD protecting the circuit to operate. This type of fault is a:

- ☐ a. Short circuit between live conductors
- ☐ b. An overload of the compressor
- ☐ c. A fault to earth
- ☐ d. Undervoltage

17. A client's car park lighting luminaires have been replaced with a different type. The lights come on at the normal time but only remain on for 30 minutes before the circuit breaker trips. The most likely type of fault is a:

- ☐ a. Short circuit between live conductors
- ☐ b. An overload of the circuit
- ☐ c. A fault to earth
- ☐ d. Undervoltage

18. A number of tungsten lights in a small office have been changed to fluorescent luminaires. When the light switch controlling these luminaires is operated there is a crackling noise. The likely cause is:

- ☐ a. Overvoltage
- ☐ b. A fault to earth
- ☐ c. Overloading of the circuit
- ☐ d. Arcing at the switch terminals

19. Excessive pressure on thermoplastic insulation against a metal accessory box can result in:

- ☐ a. A low insulation resistance fault
- ☐ b. A high insulation resistance fault
- ☐ c. A short circuit
- ☐ d. An overload

20. When working on or near live conductors safety can be improved by:

- ☐ a. The presence of a second person
- ☐ b. Taking frequent breaks
- ☐ c. Working very quickly
- ☐ d. Working alone

Diagnosing electrical faults

4

RECAP

Before you start work on this chapter, complete the exercise below to ensure that you remember what you learned earlier.

- It is important that we establish the _____ of any fault in order to determine whether a fault _____.

- The information that needs to be available for fault finding includes information relating to the events _____ up to the fault, _____' information and _____ records.

- A good _____ and understanding of the equipment, _____ and the _____ and control devices involved and the _____ of installation and _____ system is essential.

- We are responsible for _____ the building _____ wherever possible on completion of fault finding and correction.

- The common symptoms for electrical faults which we may encounter include: loss of _____, _____ voltage, operation of _____ and _____ devices, component or equipment failure or _____ and arcing.

- The maximum voltage drop within a low voltage installation supplied from a public DNO is _____ for lighting and _____ for all other circuits.

- The operation of a protective device may be caused by an _____, a _____ circuit or, in many cases, an _____ fault.

- The normal operation of switches and _____ in electrical circuits produces _____ at the terminals as the contacts _____.

- The common fault conditions which we are likely to encounter in electrical installations include _____ resistance, _____ voltages, insulation _____, excess current, _____circuit and _____ circuit.

- In electrical installations faults are likely to be found in _____ systems, _____ and _____, equipment and _____ and instrumentation and _____.

- The presence of a second person would improve the _____ of the electrician including where work is carried out on or near _____ conductors, in areas occupied by _____ or the _____, in confined or _____ access areas and in other _____ areas.

- The use of _____ and warning signs will be needed where the work involves having _____ access to holes in floors, using steps or _____, work in _____ routes and _____ and when working on or near _____ conductors or _____.

LEARNING OBJECTIVES

On completion of this chapter you should be able to:

- State the dangers of electricity in relation to the nature of fault diagnosis work

- Describe how to identify supply voltages

- Select the correct test instruments for fault diagnosis work

- Describe how to confirm test instruments are fit for purpose, functioning correctly and are correctly calibrated

- State the appropriate documentation that is required for fault diagnosis work and explain how and when it should be completed

- Explain why carrying out fault diagnosis work can have implications for customers and clients

- Specify and undertake the procedures for carrying out the following tests and their relationship to fault diagnosis:

  - Continuity

  - Insulation resistance

  - Polarity

  - Earth fault loop impedance

  - RCD operation

  - Current and voltage measurement

  - Phase sequence.

- Identify whether test results are acceptable and state the actions to take where unsatisfactory results are obtained.

Part 1 Supply voltages

This chapter considers the requirements for diagnosing electrical faults to be carried out safely.

Whilst working through this chapter you will need to refer to the IET Guidance Note 3 and the Health and Safety Executive's HSE Guidance Note GS 38, Electrical test equipment for use by electricians.

Note

The Health and Safety Guidance Note GS 38, Electrical test equipment for use by electricians is available as a free download from **www.hse.gov.uk.**

Fault finding is generally carried out during initial verification before the circuits are energized and so the electrical shock risk to the inspector is greatly reduced. When fault finding on existing installations, additional care needs to be taken as the installation and equipment will normally be energized.

We have already considered the need to ensure safe isolation when carrying out fault finding and this is essential to ensure safe working. When we begin the process of fault finding one of the key features is to determine whether there is a supply present.

To do this we need to be able to not only determine the voltage is present but also whether this voltage is appropriate and at the expected level. We have already established that reduced voltage faults do occur and can affect performance and operation of equipment.

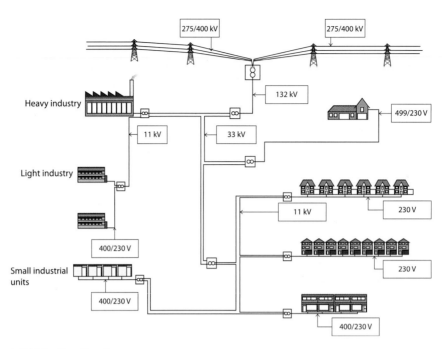

Figure 4.1 *Typical DNO distribution system*

To measure the voltage present requires a suitable voltage indicator and knowledge of the expected voltage. It is unlikely that you will encounter voltages in excess of 1000 V within most general installations. Some larger consumers purchase their energy from the DNO at high voltages (11 kV, 33 kV and 132 kV) and use their own step down transformers. Some of these consumers may even have their own HV distribution system where their installation covers a large geographical area. So, high voltages may be encountered on the consumer's distribution and transformer supply side; where such voltages exist, particular care and equipment is required and additional training is appropriate.

Within the installation there are some standard voltages which are to be expected and these can be summarized as the voltage between:

- Line conductors (U)   400 V
- Line conductors and earth (U_0)   230 V
- Line conductors and neutral   230 V.

There are items of equipment within the installation which operate at voltages other than those listed above including separated extra-low voltage (SELV), protective extra-low voltage (PELV) and functional extra-low voltage (FELV) with a maximum voltage of 50 V ac and 120 V dc.

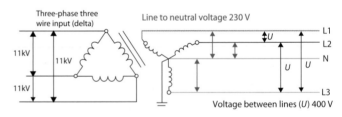

Figure 4.2 *Voltages available from an LV supply transformer*

The risk of electric shock or damage to instruments or equipment is present whenever work is carried out on an existing installation and so testing and measurement is an essential element of fault finding.

When working on electrical installations and equipment it is important to determine and confirm the voltage using an appropriate instrument. There are many instruments available on the market which are able to measure voltage, however not all of these instruments are suited to the task in hand. The Health and Safety Executive Guidance Note GS 38 provides guidance on the requirements for test instruments used by electricians.

HSE Guidance Note GS 38 identifies the risks involved in testing and these include those associated with the use of unsatisfactory test instruments. GS 38 also identifies some of the causes of accidents when testing which include unsuitable test probes, leads, lamps, voltage indicators and multimeters resulting in:

● Arcs caused by:
  ● Test probes which are not appropriately insulated
  ● Excessive current drawn through test probes and instruments due to incorrect settings.
● Electric shock caused by:
  ● Inadequate insulation of leads and probes
  ● Live terminals exposed at instruments
  ● Incorrect use of test instruments
  ● Poorly constructed/home-made test equipment
  ● Incorrect connection of leads due to poor lead identification
  ● Leads falling off and lead or connection remaining live.

In order to safely identify the voltages present it is important to use test instruments and equipment which will suitably and safely identify the voltages present.

There are some simple rules to follow when selecting a suitable voltage indicator.

It should be

● Able to indicate both ac and dc voltage
● Able to measure voltages greater than the expected supply voltage
● Confirmed to be working before and after measurement (preferably with a proving unit).

Multimeters should not be used to confirm the presence of voltage as many do not have suitable short circuit protection and can easily be left on the wrong setting. A multimeter left on the current or ohms scales will effectively be a short circuit if connected to test voltage.

a)

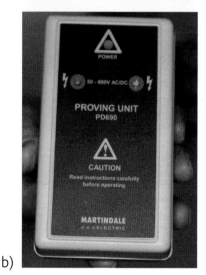
b)

Figure 4.3 *Typical approved voltage indicator and proving unit*

When carrying out fault finding the best practice approach is to use an approved voltage indicator to confirm the presence and magnitude of voltage. This is the same process as used in safe isolation as described in Chapter 2 of this study book. If a measurement of voltage is required then this will allow a voltage measurement instrument to be set or selected for the voltage indicated. Where a multimeter is used for this measurement the use of test leads which incorporate fuses are strongly recommended.

Where there is any doubt about the magnitude of the voltage to be measured using a multimeter, the highest range should be selected having first confirmed that the maximum possible voltage is within the range of the instrument.

Task

Familiarize yourself with all the requirements of HSE Guidance Note GS 38 before continuing with this chapter.

Part 2 Selection of test instruments

In addition to the confirmation and measurement of voltage there are a number of other tests which may need to be carried out when fault finding. It is important that the correct instrument is selected and that the instrument is suitable for use.

We can begin by considering the type of tests we may need to carry out and the appropriate test instruments for each type of test.

Type of test	Test instrument
Continuity of conductors	Low resistance ohmmeter
Insulation resistance	Insulation resistance ohmmeter
Polarity	Low resistance ohmmeter
Earth fault loop impedance	Earth fault loop impedance tester
Operation of RCD	RCD tester
Current measurement	Clamp-on ammeter (Tong tester)
Voltage measurement	Voltmeter or suitable multifunction test instrument
Phase sequence	Phase sequence indicator (rotating disc or indicator lamp type)

Figure 4.4 *Test instruments commonly used in fault finding*

These test instruments are available in a number of formats, from basic individual instruments for each test, to the multifunctional test instrument which combines a number of test functions in a single instrument.

> **Remember**
>
> **Multifunction test instruments should not be used to confirm the presence of a voltage. Once presence and magnitude are established actual measurements may then be made.**

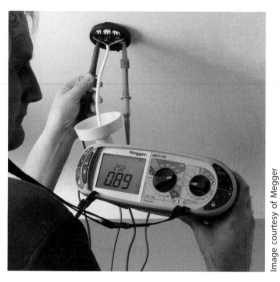

Image courtesy of Megger

Figure 4.5 *Typical multifunction test instrument*

When a multifunction test instrument is to be used the instrument must be set to the appropriate function for the test which is to be undertaken.

> **Remember**
>
> **You must take particular care to ensure multifunction test instruments are set correctly before connection to the circuit or equipment being tested.**

We must also consider the requirements for the use and suitability of the instruments to be used and the first requirements relate to our competence and so we must:

- Comply with the requirements of the Electricity at Work Regulations
- Understand the test equipment being used, including its setting and rating
- Understand the characteristics of the installation or equipment being tested.

We must then carry out some basic checks on the instruments we are to use to confirm they are suitable and safe to use. We are going to rely on the results of the tests to enable us to confirm the presence and location of the fault. It is important therefore to confirm that the test instruments are going to provide accurate results.

The first check is that the test instrument complies with the appropriate British Standard which is generally BS EN 61010. Earlier instruments could have been manufactured to BS 5458 which has been withdrawn but providing they meet the current specification they are suitable for use.

Providing the instrument meets the appropriate British Standard the next check is to confirm that the calibration is in date. This process of calibration checks the instrument accuracy against a national standard and confirms that, within the manufacturer's tolerances, the instrument is giving accurate readings. The instrument should be regularly calibrated in accordance with the manufacturer's instructions which, in most cases, is annually.

CALIBRATION

COMMENTS _____

_____

BY _____   DATE _____

NEXT CALIBRATION _____

Figure 4.6 *Typical calibration label*

The test instruments should be checked regularly to make sure the readings continue to be accurate between calibrations. The process is known as ongoing accuracy and is necessary because test instruments are carried from site to site in a vehicle, subjected to all sorts of environmental conditions, vibration and possible impact. The accuracy of the instruments can be affected by the environment and their use in fault finding. Checks done regularly, say once a month, confirm continuing accuracy and the results of the checks are recorded.

Many test instruments rely on internal batteries for their power source. There are two further checks that need to be carried out on the instrument before it is put to use.

Battery condition is important for the instrument to function correctly and give accurate results. Before the instrument is used we need to confirm that the battery has a suitable output voltage level. How this is done and what output voltage is acceptable will vary from one manufacturer to another. The accuracy of the instrument is dependent upon this battery voltage and so it should always be checked before testing begins.

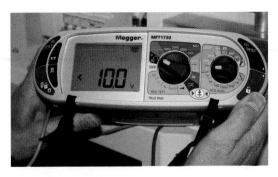

Figure 4.7 *Battery voltage check*

Where it is appropriate we must also confirm that the instrument is *functioning correctly* before carrying out tests. This may be done by, for example, operating the test instrument with the test leads connected together (a low resistance expected) and with the test leads apart (a high resistance expected). For test instruments which rely on an external supply for their operation, their function cannot normally be confirmed before carrying out tests on site.

Finally we need to confirm that the test leads, probes and clips are in good condition and not damaged and that they and the instrument comply with the requirements of HSE Guidance GS 38.

Try this

1   Identify the test instrument to be used when investigating the following faults on a lighting circuit.

   a   line conductor open circuit

   _____

   _____

   b   short circuit between live conductors

   _____

   _____

c suspected overload

_____

_____

d suspected undervoltage.

_____

_____

2 A low resistance ohmmeter is to be used to locate a circuit fault. List **four** checks that need to be made on the instrument, **not** including the leads, to confirm it is suitable to carry out the tests.

a _____

b _____

c _____

d _____

Part 3 Implications for the client and documentation

Implications

Before we look at fault finding testing it is important to understand the implications of the work for the client or customer. It is quite likely that the client has already been inconvenienced as a result of the fault and we need to try and minimize any further inconvenience.

As stated earlier in this study book the fault diagnosis requires information from the client to allow the work to progress safely and efficiently. The client needs to make this information available to us before the work can begin. In some complex installations permits to work, method statements and a safety policy may be required.

In some locations induction training may be necessary for anyone visiting or working within the premises.

These requirements will have an impact on the client or customer insomuch as some information and/or training needs to be provided by them; other information needs to be provided to and approved by them. All of which can involve considerable client resource and increase the time and cost for the whole fault finding and repair process.

As fault finding is, in most cases, an unknown quantity it is not easy to give the client a clear indication of the time involved in the location and rectification process. We need to gather information from staff involved with the installation or equipment and this will affect their productivity. The fault finding process may involve access to a number of areas and could involve interrupting the client's work processes, often for an unspecified period of time.

In some instances the need to locate and repair the fault is essential and will take precedence over everything else. The loss of ventilation to a battery farming facility, for example, will require urgent attention as the rise in temperature and distress to the livestock will be rapid and could quickly prove fatal. It is worth remembering that under these conditions there will be a lot of pressure on the electrician to complete the work quickly. This should never lead you to take short cuts and risk the safety of yourself and others, including livestock, whilst carrying out the work.

Figure 4.8 *Ventilation is essential for battery farms*

On other occasions the location and repair may be put off to an out-of-working hours period because the disruption caused outweighs the particular problem caused by the fault. For example, one finishing machine is not working and is on a circuit and in a location with the remaining five machines. To avoid the disruption to production and associated cost, the location and repair of the fault may be put off until the production run is completed.

Figure 4.9 *Fault finding in some locations may cause considerable disruption*

During the fault finding and repair process part of the installation or equipment will be out of service and this may extend beyond that originally affected. For example, a defective circuit breaker or RCD would require the distribution board to be isolated to effect a replacement. This could seriously disrupt the client's production and it may need to be carried out outside of normal working hours.

The cost of replacement parts and equipment may be considerably less than the cost in production for manufacturing companies, and office and retail facilities also stand to lose custom or data. In each case the additional cost in man hours may be considerable not only for the client or customer but also for the employees where they are paid production bonuses.

Documentation

When carrying out fault finding we need to keep a record of the information and test results which are obtained during the process and the subsequent rectification. Many companies use their own forms for recording test results and measurements. It is common for electricians to produce a rough layout sketch and mark the

Charles Machinists Ltd		Repair record				No. 00173
Date	Equipment Ref No	Fault	Symptoms	Cause	Material	Re-commissioned
--/07/20--	Workshop heater WH1	No heat and fan boost not operating	Operating OK and then stopped working	Traced to thermal cut-out fused	Replacement thermal cut-out TK4-R120	--/07/20—am.
Name: James Douglas		Signature: *James Douglas*				Date: --/07/20--

Figure 4.10 *Repair record for equipment repair*

location of equipment on the layout. Test results and findings are then noted on the sketch wherever these are undertaken. The results of this may be transferred to a record sheet to produce a test record for the process. Often the sketch is copied and kept in the job file for reference.

An overall report is then produced, similar to the one we considered earlier in this study book. This gives the client a record of the fault, the cause and rectification for their records, and enables them to demonstrate compliance with their statutory health and safety obligations. Where the work involves fault finding and repair on a specific item of equipment this record is normally all that is required.

Where the work involves the fixed wiring of the installation then a certificate needs to be issued to confirm the compliance of the work with BS 7671 and to record any observations on the existing installation.

Where a single circuit is involved then a Minor Electrical Installation Work Certificate (MEIWC) is suitable. Where a number of circuits are involved then an Electrical Installation Certificate (EIC) may be more appropriate than a number of MEIWCs.

Figure 4.11 *MEIWC for the rectification of the fault*

The section for comments on the existing installation is quite important where fault finding is concerned. It may be that the rectification of the fault is a priority and may be necessary to remove an immediate danger. It is not necessary for the whole installation to be brought to the current standards but any non-compliance that we observe on the existing installation must be recorded. This notifies the client of their existence and makes them aware of the need for improvement.

It may be suitable to use a copy of the schedule of test results, which can accompany the EIC to record tests carried out during the fault location process. This can be used with the location recorded rather than the circuit, and the test results at each location recorded in the appropriate column. This would not form part of the electrical installation certificate as that schedule relates to circuits and is used to confirm their compliance, which may not be the case during the fault location process.

out due to site restrictions. For example, it may not be possible to carry out an insulation resistance test between live conductors because of the connected equipment. The work carried out would be fully tested before it is reconnected to the circuit, but the confirmation of the circuit compliance would require all the equipment to be disconnected. In this instance the L-L value would be recorded as LIM (limitation) and with the live conductors connected together a test to L&N to earth would be carried out and recorded.

The tests are carried out to enable the electrician to confirm the circuit complies with BS 7671 and is safe to put back into service following the repair.

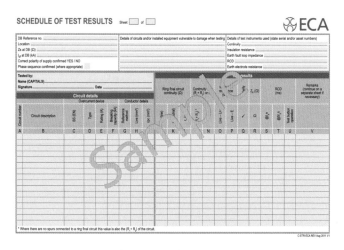

Figure 4.12 *Schedule of test results*

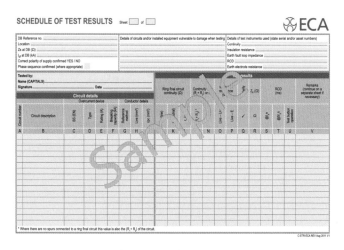

Figure 4.13 *Essential tests*

> **Remember**
>
> **The work you undertake must meet the current requirements of BS 7671, any other non-compliances you observe must be recorded on the certification. You are not required to look for non-compliances but any you see during the course of your work must be recorded.**

We have considered documentation used:

- During the fault location and repair
- To record what action was taken
- To confirm the installation or equipment is safe to be placed in service.

The documentation recording the fault, cause and remedial action, together with the appropriate electrical installation certification, should be issued to the client on completion of the work.

When working on a single circuit the MEIWC identifies the essential tests which are carried out to confirm compliance. There may be occasions where some of these tests may not be carried

Copies of all the documentation produced during the whole process should be copied and retained on the contractor's job file.

Try this

1   List the documents that would be completed for recording:

   a   the test results during the fault finding process

   _____

   b   the symptoms, cause and remedy of a fault

   _____

   c   that a circuit is safe to put back in service.

   _____

2   A fault has been reported on a circuit within a complex manufacturing facility with some specialized processes. List three requirements which may have implications for the client before the fault finding process can begin.

   _____

   _____

   _____

   _____

   _____

   _____

   _____

   _____

Part 4 Testing for fault diagnosis

Many of the tests we carry out during fault diagnosis are similar to those used for testing installations during initial verification. The application of the tests and identifying the location of the fault requires a logical approach. This will vary depending on the nature of the circuit or equipment and the symptoms of the fault. We will consider the most common tests related to the installation and the methods used in locating the fault.

Note

More detailed Information on the tests considered here for fault finding are contained in the *Inspection Testing and Certification* study book in this series.

Continuity

As we are measuring conductor resistance this will involve low values of resistance-that being the main feature of a conductor. The test instrument to be used for this test is a low resistance ohmmeter and this has specific characteristics which include:

- An output voltage between 4 V and 24 V
- A short circuit current of 200 mA
- A resolution of 0.01 Ω.

Note

The characteristics of the instruments used to carry out tests on electrical installations are given in IET Guidance Note 3, Section 4.

Because the values of resistance are expected to be low, the effect of the resistance of the connections and leads can be significant. The inspector must compensate for this resistance. This may be done by either:

- Measuring the resistance of the test leads and subtracting this resistance from the measured values
or
- Using the null facility on the test instrument where this is provided, to subtract the test lead resistance and so give true values direct from the instrument.

Most digital test instruments have lead null facility.

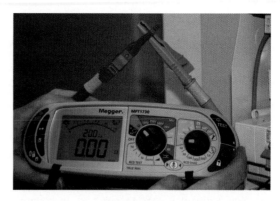

Figure 4.14 *Low resistance ohmmeter with test leads nulled*

Two things to remember when measuring or nulling the test leads are to ensure that:

- The test leads are fitted with the crocodile clips
- The clips are joined together with the fixed jaws together **NOT** the sprung jaws.

The sprung jaws are only connected to the leads by the spring within the crocodile clip and this will add resistance which is not present when the solid jaw is clamped against the conductor. If the leads are incorrectly nulled then too much resistance will be automatically removed from the reading and the test results will not then be accurate.

There are two basic methods of testing to confirm the continuity of conductors.

Method 1

This test is the same as the test generally referred to as the $R_1 + R_2$ test, commonly used to confirm the continuity of the circuit protective conductors at initial verification. The term '$R_1 + R_2$ test' comes from the fact that it involves linking the line (R_1) and cpc (R_2) together at the distribution board and measuring between the line and cpc at each point on the circuit.

This test can be carried out using any two conductors when fault finding and is useful when looking for open circuit faults.

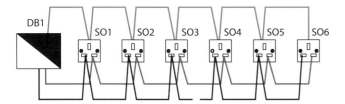

Figure 4.15 *Radial circuit*

Example: Figure 4.15 shows a radial circuit supplying six socket outlets and on checking for presence of supply we find there is no supply to the socket outlets 4, 5 and 6. We can see there is a break in the neutral between socket outlets but on site this would not be visible. If the circuit is protected by an RCD it may not be possible to test using a voltage indicator.

By isolating the circuit and linking line and cpc at the distribution board (DB) then, using a low resistance ohmmeter, testing between line and earth at each socket outlet would confirm the line and cpc are continuous throughout. We then link line and neutral at the DB and test line to neutral at each socket outlet. Continuity would be proved up to SO3 and then open circuit at SO4 indicating the break is between SO3 and SO4.

Method 2

An alternative method is to use a long or wander lead and test between each point on the circuit to confirm continuity. The long test lead should be nulled out before the testing is undertaken and care needs to be taken to ensure the lead does not become a trip hazard.

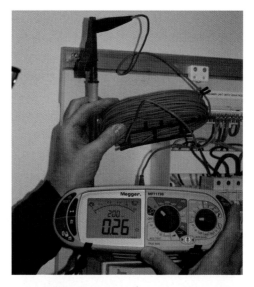

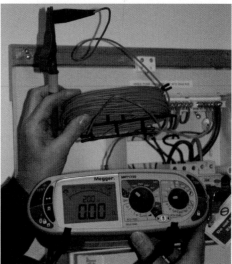

Figure 4.16 *Long lead nulled*

We can minimize the length of the test lead by testing from the DB to SO1 and then from SO1 to SO2 and so on. Of course on site the points on the circuit may be some distance apart and often in different rooms.

We can use the test results to establish whether the circuit is healthy by using standard values for conductor resistance. IET Guidance Note 3 contains details of conductor resistances in mΩ per metre in Appendix B Table B1.

Task

Familiarize yourself with Table B1 in IET Guidance Note 3, Inspection and Testing before continuing with this chapter.

The table contains generic details on the conductor resistances and this will allow us to confirm the results are acceptable. If we know the length of the circuit being tested, the expected resistance can be calculated and the measured resistance compared with this expected value.

Example: The radial circuit in Figure 4.15 has been installed using 2.5 mm² flat profile twin and cpc cable and is 50 m long to the furthest point. To determine the expected line to cpc resistance we need to use the information from the table and the conductor length. As the cable

$R_1 + R_2$ resistance in mΩ/m at 20 °C		
csa in mm²		Resistance in mΩ/m
Line	cpc	Copper
2.5	–	7.41
2.5	1.0	25.51
2.5	1.5	19.51
2.5	2.5	14.82

Figure 4.17 *Conductor resistance values for 2.5 mm² line conductors and their cpcs*

is 2.5 mm² flat profile, the cpc will be 1.5 mm² and so using the information in Figure 4.17 the $R_1 + R_2$ resistance per metre will be 19.51 mΩ/m. The expected line to cpc resistance for the circuit will be:

$$R_1 + R_2 = \frac{m\Omega m \times L}{1000} = \frac{19.51 \times 50}{1000} = 0.9755\,\Omega.$$

The instrument is required to have a resolution of ≥ 0.01 and so the displayed figure would be in

the range 0.97 Ω to 0.98 Ω, depending on the manufacturer's tolerance settings and how the values are rounded up or down.

Faults may arise due to poor connections or terminations in a circuit which will result in a high resistance. Knowing the expected resistance values per metre we can check using either method to confirm the values are within the expected parameters.

Where we find that there is a problem with the circuit the normal first step would be to go to roughly the midpoint and disconnect. We can then apply the tests to both halves of the circuit and so narrow down the location of the fault.

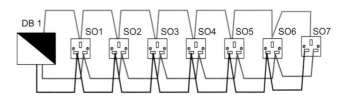

Figure 4.18 *Radial circuit*

For example, in Figure 4.18 disconnecting the circuit at SO4 allows us to test L-N from the DB to SO3 first. Placing a link between the line and neutral at SO7 would allow us to test L-N for SO5 to SO7 and so on.

This process would need to be followed for a ring final circuit as each point is effectively supplied from both directions. To locate a fault it is common to disconnect the ring final circuit from the distribution board and at or close to the midpoint and then test each half as a radial circuit. Each section can then be further broken down to locate the fault in the same way as we did for the radial circuit.

We can see from these examples how the measurement of conductor resistance can provide valuable information regarding the condition of the circuit and help with fault finding.

Note

Where short lengths of conductor are involved, for example 2.5 m of 10 mm^2 copper conductor, then the resistance will be very low ($2.5 \times 0.00183 = 0.0045\ \Omega$). This may be displayed on the instrument as 0.00 Ω due to the instrument resolution being 0.01 Ω. Continuity will have been confirmed and the result to be recorded would be < 0.01 Ω showing it was less than the instrument could measure.

Remember

Where circuits are installed in steel containment systems the steel containment will create parallel paths which will result in the measured cpc values being lower than expected.

Try this

A radial circuit supplying a number of points has a total length of 30 m and is installed using thermoplastic insulated 2.5 mm^2 copper conductors installed in PVC conduit. The line and neutral conductors have been linked at the DB and a test line and neutral at the furthest point of the circuit produces a reading of 0.6 Ω. Determine, using the information in Figure 4.17, whether this is an acceptable value.

_____

_____

_____

Insulation resistance

When fault finding on electrical installations the test for insulation resistance is used to identify and locate low resistance faults and short circuits between live conductors and live conductors and earth.

Insulation resistance testing is used to measure the resistance of the insulation separating live parts-conductors in particular, from both each other and the earth. So we are looking for high values of resistance and the instrument used, an insulation resistance ohmmeter, must be capable of measuring resistance in the MΩ range.

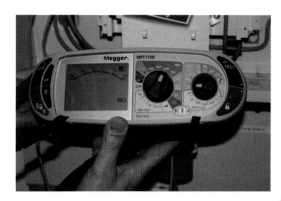

Figure 4.19 _Insulation resistance ohmmeter_

Before carrying out the tests, the instrument checks must be carried out to confirm that it is safe and suitable for use and the test leads comply with the requirements of GS 38.

The test is carried out using a voltage which is generally higher than that to which the circuit will normally operate. The applied test voltage is dependent upon the normal operating voltage of the circuit to be tested.

BS 7671 identifies the minimum acceptable insulation resistance values and the applied test voltages for installations (not circuits). When we are fault finding we would expect a healthy circuit to have values far higher than these minimum values.

Nominal circuit voltage	Test voltage (dc)	Minimum insulation resistance
SELV and PELV only	250 V	0.5 MΩ
Up to and including 500 V (excluding SELV and PELV only)	500 V	1.0 MΩ
Over 500 V	1000 V	1.0 MΩ

Figure 4.20 *Insulation resistance test applied voltages and minimum acceptable values*

Faulty circuits may have either a low resistance fault which is below the minimum acceptable value or a short circuit which will give 0 MΩ. One of the factors which affects insulation resistance is the length of the cable. The resistance is inversely proportional so the longer the cable the lower the insulation resistance.

There are two main safety considerations when carrying out an insulation resistance test:

● The circuit must be isolated from the supply
● The test voltages being injected into the installation are high and could cause danger.

Whilst the output is unlikely to result in a fatal electric shock the involuntary reaction to the shock can result in falls from height and other physical injuries.

Before we can carry out an insulation resistance test on a circuit there are a number of procedures that must be carried out including:

● Ensuring the supply is securely isolated and locked off and that there is no supply to the installation or circuits to be tested
● Removing all loads
● Disconnecting all equipment that would normally be in use
● Disconnecting and bypassing (linking out) any electronic equipment that would be damaged by the high voltage test (this includes dimmer switches, photocells, passive infrared sensors, electronic controllers and control gear and certain RCDs)
● Ensuring there are no connections between any line or neutral conductor and Earth
● Any downstream cbs are switched on and all fuses in place
● Putting all control switches in the on position (unless they protect equipment that cannot otherwise be disconnected)
● Testing two-way or two-way-and-intermediate switched circuits with the switches in each direction (unless they are bridged across during the testing).

The test is then carried out on the isolated circuit between all the live conductors and all live conductors and Earth. For a healthy new circuit in good condition it is quite possible that the reading will indicate the resistance is over the range of the instrument. If the maximum range on the instrument is say 999 MΩ then the result that would be recorded would be > 999 MΩ showing it was above the range of the instrument.

When fault finding it is important to ensure that all the loads and equipment are disconnected or isolated from the circuit. Failure to do so may damage the equipment and give misleading results. It is important to talk to the client and user of the installation to establish where all the connected equipment is located.

For example:

A TV signal amplifier in the roof space is left connected in an upstairs socket outlet circuit where the insulation resistance is to be tested. This results in a measured value of 0.00 MΩ L-N and the amplifier may be damaged as a result of the applied test voltage. It could also result in several hours of time, considerable effort and disruption to find that the circuit is in fact healthy and that the load had been overlooked and left in circuit.

A neon indicator on an unused fused spur connection unit is left in circuit. When the circuit is tested line to neutral a value of around 0.24 MΩ is recorded. A number of neon indicators left in circuit will produce the same effect as resistors connected in parallel and bring the value down even further. Whilst this test will not damage the neon indicator it can result in wasted time, effort and disruption whilst the cause is located.

Note

It is advisable to first perform a test L-N at the lower voltage, 250 V dc, when fault finding. If this results in a very low or 0 MΩ reading investigate further to establish whether there is still equipment connected to the circuit.

When testing three phase and three phase and neutral circuits, the test should be carried out between all live conductors and all live conductors and Earth, just as we did for the single phase installation. With these circuits the tests between

Try this

An insulation resistance test is to be carried out on a circuit supplying a number of fused spur outlets with neon indicators mounted above a workbench and supplying soldering irons. List five procedures you will need to carry out to prepare the circuit for testing.

1 _____

2 _____

3 _____

4 _____

5 _____

live conductors is more involved as we have to test between each line and every other line, and between each line and neutral. Tests are also required between each live conductor and Earth.

In all there are a total of ten test stages but this can be reduced by, for example, linking all the line conductors together and testing them all to neutral at once. Providing the result is satisfactory then individual line to neutral tests would not be necessary. If a fault is identified then each line must then be tested to neutral individually, to enable the fault to be located.

Polarity

The polarity of a circuit may be confirmed using the tests we carried out when testing continuity.

The $R_1 + R_2$ test carried out linking the line and cpc, then testing at each point on the circuit can be used to confirm these conductors are correctly connected. The long lead test can also be used, only on this occasion the lead is connected to the line conductor and the test is to each line connection on the circuit.

Tests would also be required to confirm that the neutral and cpc connections are correctly connected.

Confirming polarity is carried out to ensure that:

● All single pole protective and control devices are connected in the line conductors only
● The centre pin of each Edison screw (ES) lampholder is connected to the line conductor only
● All equipment and socket outlets are correctly connected.

If the protective device for the circuit is not connected in the line conductor then the circuit and equipment:

● Will not be isolated when the switch or circuit breaker is off or the fuse removed and so the circuit cannot be made safe
● If the protective device is used for fault protection and is connected in the neutral conductor then a fault to earth will not be detected and the protective device would not operate. The simple earth fault path in Figure 4.21 shows that the fault current to earth will not flow through the protective device.

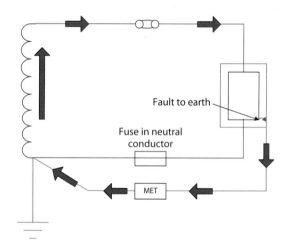

Figure 4.21 *The earth fault path with the fuse in the neutral conductor*

For a circuit that has been energized and in use we may confirm polarity by carrying out a test using an AVI.

Figure 4.22 *Approved voltage indicator*

The AVI must meet all the requirements of HSE Guidance GS 38. The check for polarity is carried out at every relevant point on the circuit between:

- All live conductors
- All live conductors and Earth.

For a single phase installation this is between line and neutral, line and Earth, and neutral and Earth. On a three phase installation the check is carried out between line $(L)_1$ and L_2, L_1 and L_3, L_2 and L_3, L_1 and N, L_2 and N, L_3 and N, L_1 and E, L_2 and E, L_3 and E, and N and E.

For a three phase 400/230 V supply this should be between:

- L_1 and L_2, L_1 and L_3, L_2 and L_3 = 400 V
- L_1 and N, L_2 and N, L_3 and N = 230 V
- L_1 and E, L_2 and E, L_3 and E = 230 V
- N and E = 0 V.

For a single phase 230 V supply the result should be between:

- L and N = 230 V
- L and E = 230 V
- N and E = 0 V.

Earth fault loop impedance testing

The earth fault loop impedance test can be considered in two separate stages:

- external earth fault loop impedance (Z_e)
- system earth fault loop impedance (Z_s).

External earth fault loop impedance (Z_e): This test is to confirm that the external earth fault path is present from the electrical installation to the source earth of the system, and that the value is within the defined limits. If there is no return path to earth then there is a considerable risk of electric shock.

Remember

The DNO is not obliged to provide an earth connection for the consumer. If they do not provide an earth connection the consumer is responsible for providing the earth for their installation and the supply will be TT.

The test instrument used is the earth fault loop impedance tester.

Try this

State the:

a Two test methods that are used to test for polarity

_____

_____

b Effect on the operation of a circuit breaker connected in the neutral conductor only in the event of a fault to earth.

_____

_____

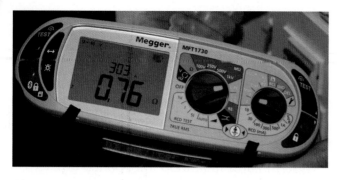

Figure 4.23 *Instrument set to measure Z_e*

As the earthing conductor is to be disconnected for this test, permission will be required from the responsible person before the installation is isolated for this test.

Procedure for the test to confirm Z_e:

- seek permission to isolate
- safely isolate and lock off the supply
- disconnect the earthing conductor from the main earthing terminal (MET)
- select an earth fault loop impedance tester
- confirm instrument is in calibration and not damaged
- confirm test leads comply with GS 38

- connect test instrument to:
    - disconnected earthing conductor
    - incoming line terminal of the main isolator
- test and record the result
- reconnect the earthing conductor to the MET
- confirm test result is compliant.

The result is compared to the maximum acceptable value for the external earth fault loop path for the particular supply system.

The DNO provides details of the maximum earth fault loop values for the public distribution network as:

- TN-S 0.8 Ω
- TN-C-S 0.35 Ω.

When testing Z_e the expected result should be ≤ the stated values.

If the system of supply is TT then, in all but very exceptional circumstances, an RCD will be required to provide earth fault protection. The value of earth fault loop impedance for TT systems is dependent upon the rating of the RCD used. Figure 4.25 gives the maximum values which are acceptable.

Maximum external earth fault loop impedance values for TT systems	
RCD $I_{\Delta n}$	Maximum R_A in Ω
30 mA	1667 Ω*
100 mA	500 Ω*
300 mA	167 Ω
500 mA	100 Ω

* BS 7671 advises that values in excess of 200 Ω may prove to be unreliable.

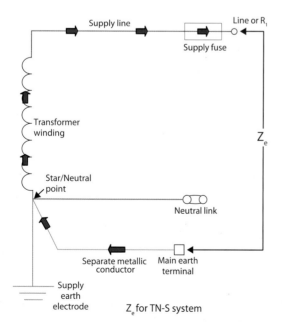

Figure 4.24 *Z_e circuit for a TN-S system*

Figure 4.25 *Maximum earth fault loop impedance values for RCD protected installations*

It may be necessary to carry out a test of Z_e in the event of the failure of the installation protective devices to operate under earth fault conditions. This will normally be apparent when further damage has occurred to equipment or conductors.

System earth fault loop impedance (Z_s)

The earth fault loop impedance Z_s has to be low enough to allow the protective device to disconnect within the required time. In the event of a fault to earth, the fault path consists of the circuit line conductor and cpc ($R_1 + R_2$) and the external earth fault loop path (Z_e), so we can determine the expected Z_s value for a circuit using the formula $Z_s = Z_e + (R_1 + R_2)$.

The test is carried out with the conductors at a temperature considerably lower than their normal operating temperature as they are not in service or carrying the full load current.

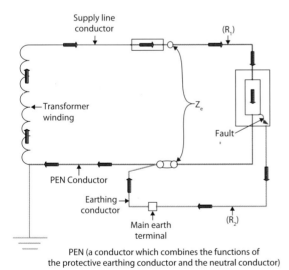

Figure 4.26 *TN-C-S earth fault loop path* {$Z_s = Z_e + (R_1 + R_2)$}

BS 7671 provides maximum earth fault loop impedance values for the different types of protective device and disconnection times. These tabulated values are at the normal operating temperature of the conductors (70 °C or 90 °C etc.). The conductor temperature at the time of the test will be lower and so we must compensate for this temperature difference.

Remember

The resistance of conductors used in electrical installations increase as temperature increases. As current increases so temperature increases and it may take several hours for a conductor to reach its normal operating temperature. It is at normal operating temperature that the conductor resistance will be at its highest.

One method of compensating for the difference between the conductor temperature at the time of testing and the normal operating temperature of the conductor, is to apply a multiplier of 0.8 to the BS 7671 tabulated values.

Task

Familiarize yourself with Tables 41.2 to 41.4 in BS 7671 before continuing with this chapter.

Example: A circuit is protected by a 16 A BS EN 60898, Type B cb. The maximum tabulated value of earth fault loop impedance given in BS 7671 is 2.87 Ω. As this is at the operating temperature of the conductor we must compensate for the difference between the normal operating temperature and the conductor temperature at the time of test. We use the 0.8 multiplier (referred to as the rule of thumb) and so the maximum measured $Z_s = 2.87 \times 0.8 = 2.296$ Ω.

Alternatively, we can use the figures given in the IET Guidance Note 3, Inspection and Testing or the IET On-Site Guide which have already been temperature corrected and give the maximum measured Z_s.

The test is carried out with the circuit energized and so will require access to live terminals at the furthest point on radial circuits. An additional test lead fitted with a 13 A plug is available to allow this test to be carried out on ring and radial socket outlet circuits. Where this lead is not available, the use of a plug-in adapter is very beneficial when testing socket outlets.

The test is carried out using the earth fault loop impedance tester. Many instruments provide two options for carrying out the test. The test is carried out between line and Earth at:

- A current which may be as high as 25 A (often referred to as the high current test)
- A low current or 'no trip' test which uses a much lower current (mA) and is used, for example, where the circuit is protected by an RCD.

A problem may occur when testing at the higher current with some cbs, such as a BS EN 60898 6 A, Type B as the high current Test may be over four times the rating of the device. In such cases the value of Z_s can be determined by calculation using $Z_s = Z_e + (R_1 + R_2)$.

Remember

The measured values of both Z_e and $R_1 + R_2$ must be used when determining Z_s.

The alternative low current or no trip option generally provides a less accurate result. This is because of interference from harmonics on the system and the use of very low current using repeated sampling.

Using either of these methods to carry out an earth fault loop impedance test on a radial circuit supplying 13 A socket outlets, the process will be the same:

- Locate the furthest socket outlet on the circuit
- Use a plug-in adapter or proprietary test lead
- Select an earth fault loop impedance tester
- Confirm instrument is in calibration and not damaged
- Confirm test leads comply with GS 38
- Connect test instrument
- Test and note the result
- Unplug the test instrument
- Confirm test result is compliant
- Record the result.

Try this

A 40 A BS EN 60898, Type B circuit breaker protects a power circuit. The maximum tabulated value of earth fault loop for this circuit is 1.15 Ω. The measured earth fault loop impedance at the equipment supplied on this circuit is 0.85 Ω. Determine whether the circuit meets the requirements for fault protection.

_____

_____

_____

_____

When investigating faults in an electrical installation it may be necessary to carry out the earth fault loop impedance test at other points on the circuit. It may be necessary to carry out a test of Z_s in the event of:

● Failure of a circuit protective device to operate under earth fault condition

● On completion of remedial work to ensure the circuit complies with the requirements for fault protection.

Testing residual current devices

RCDs are installed for a number of reasons:

● Additional protection

● Earth fault protection

● Where there is a risk of fire.

RCDs include an integral test button which should be operated quarterly by the user of the installation to confirm the operation of the mechanism. Electrical tests are carried out to confirm an RCD operates within the Standard parameters. The RCD can be tested at its output terminals which is often the easiest place to test where the RCD protects any circuit other than a socket outlet circuit.

RCDs need testing in order to confirm their correct operation. The tests which need to be carried out on each device are:

● Simulation of a fault by using an RCD test instrument

● Operation when the integral test button is operated.

It is important that the electrical fault simulation tests are carried out before the test button is checked as this may affect the performance of the device.

BS 7671 identifies specific installations and requirements for RCDs in two categories:

● Additional protection: using RCDs with $I_{\Delta n} \leq$ 30 mA in installations or locations where there is an increased risk of electric shock

● Fault or fire protection: protection against electric shock where a suitable earth fault loop impedance cannot be achieved (most TT installations) or where there is an increased risk of fire.

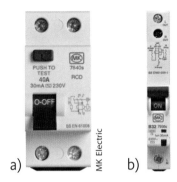

Figure 4.27 *Typical RCD and RCBO*

> **Note**
>
> Before testing RCDs we must confirm there is a suitable earth fault path and so an earth fault loop impedance test must be carried out first. This may be done from the incoming terminals of the RCD.

An RCD test instrument is used for the test and needs to supply a range of test currents appropriate to the device being tested, plus display the time taken for the device to operate. RCD test instruments have the facility to carry out the test in both the positive and negative half cycles of the supply. Each of the tests should be carried out in both half cycles and the highest value obtained is recorded.

Additional protection

We shall start by considering the tests to be carried out on a 30 mA RCD, installed to provide additional protection; the sequence of tests given here is the most appropriate. In all there are three tests to be carried out and each on both the positive and negative half cycles of the supply. Each test applies a test current based upon $I_{\Delta n}$ of the RCD and the RCD should operate within the prescribed time in each case. As the RCD will operate (trip) whilst the tests are carried out, we need to obtain permission to isolate the circuit(s) or installation which it protects.

The RCD tester is checked to confirm that calibration is in date, that it is not damaged and that the leads are compliant with GS 38. The test instrument is then set to the operating current $I_{\Delta n}$ of the RCD so in this case 30 mA.

Using a two lead test instrument and testing at the outgoing terminals of the RCD, the test instrument is connected to the earth bar and the outgoing line of the RCD. With the RCD switched on, the test is carried out and the RCD should trip, with the time displayed on the screen of the instrument. Note the time shown.

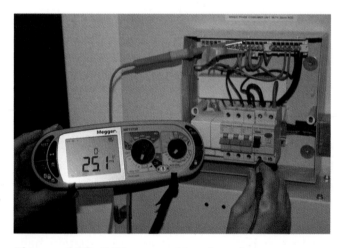

Figure 4.28 *RCD test at 30 mA and a test current of 5 × $I_{\Delta n}$*

The process for testing a 30 mA RCD installed for additional protection is:

- Obtain permission to isolate (the device will switch off)
- Select an RCD tester
- Confirm instrument is in calibration and not damaged
- Confirm test leads comply with GS 38
- Set the instrument to the 30 mA setting
- Set the test current to 150 mA (5 × $I_{\Delta n}$)
- Connect test instrument to the MET and the outgoing line terminal of the RCD
- With the RCD switched on test at 0° and 180° resetting after each operation
- Record the highest of the two results as 5 × $I_{\Delta n}$ value
- Set the test current to 30 mA (1 × $I_{\Delta n}$)
- Repeat the test at 0° and 180° resetting after each operation
- Record the highest of the two values as the 1 × $I_{\Delta n}$ value
- Set the test current to 15 mA (½ × $I_{\Delta n}$)
- Repeat the test at 0° and 180° where possible
- Disconnect the test instrument
- Operate the RCD integral test button
- Reinstate the supply.

Note

The five times $I_{\Delta n}$ test is only carried out on RCDs ≤ 30 mA which are used for additional protection.

RCD operating times		
Test current	Operating time	Time delay or S type
5 x $I_{\Delta n}$	40 ms	
1 x $I_{\Delta n}$ (BS 4293)	200 ms	200 ms + 50 to 100% of delay
1 x $I_{\Delta n}$ (BS EN 61008 & 9)	300 ms	130–500 ms (S type)
½ x $I_{\Delta n}$	No trip	

Figure 4.29 *Operating times for RCDs*

RCD test sequence

There is no set sequence for the RCD tests and some test instruments have an auto test which, once selected and started, carries out the required tests automatically, allowing the electrician to be free to reset the RCD at each stage.

This may be useful when testing from a socket outlet on a circuit but is not really necessary when the test is performed from the output terminals of the RCD. This automatic test sequence is generally $\frac{1}{2} \times I_{\Delta n}$, $1 \times I_{\Delta n}$ and $5 \times I_{\Delta n}$.

The $\frac{1}{2} \times I_{\Delta n}$ test is to establish whether the device is likely to suffer from nuisance tripping under normal operation. If the test is carried out as the first test there is no assurance that the device has not tripped simply because the mechanism is sticking.

If the $\frac{1}{2} \times I_{\Delta n}$ test is carried out as the final test, the RCD will have operated at least four times so if it is going to nuisance trip it will do so then.

RCDs installed for fault or fire protection

RCDs with an $I_{\Delta n} > 30$ mA are often installed to provide fault protection where suitable earth fault loop impedance cannot be achieved (most TT installations) or where there is an increased risk of fire. These devices will also need to be tested to ensure they are performing to the required standard. The test process for these RCDs is exactly the same **except** the $5 \times I_{\Delta n}$ test is **not** carried out- In which case only the $1 \times I_{\Delta n}$ test and the $\frac{1}{2} \times I_{\Delta n}$ test are required.

RCDs connected in series

In some circumstances it is necessary to install RCDs in series with one another, for example where a distribution cable and final circuits are connected to a TT system.

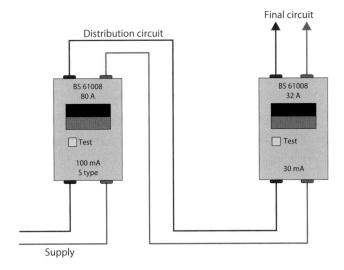

Figure 4.30 *RCDs connected in series*

The correct operation of these RCDs must be achieved, as a fault on the final circuit should not cause the RCD upstream to operate first. To ensure the correct operation of the devices and minimize disruption, a time delay or S type RCD is installed upstream. These S type RCDs are manufactured to BS EN 61008 and BS EN 61009 and can be adjusted between 130 ms and 500 ms. This allows the operation of the upstream RCD to be delayed and the downstream RCD to operate first. If the downstream RCD fails to operate then the upstream RCD will still disconnect the supply after the delay period.

Earlier RCDs to BS 4293 are time delay RCDs and the operating time of these devices can also be adjusted to allow the correct operation of the RCDs.

One of the common fault conditions reported in relation to RCDs is nuisance tripping, where the RCD operates randomly and can be reset immediately. This type of fault may be caused by:

● Oversensitive RCD tripping at a lower than acceptable value
● Normal operation of equipment causing an imbalance between line and neutral. e.g. inductive loads

- Possible fault on the installation, although this often prevents the RCD being reset
- Combined circuit leakage exceeding the operating current of the RCD, although again this often prevents the RCD being reset or it tripping quickly after resetting unless equipment is switched off.

There are a number of possibilities here and so we need to follow a logical approach to locate the problem.

Ramp testing

Many RCD test instruments include a ramp test setting which is particularly useful when determining whether it is the RCD at fault. At this setting, the test instrument applies the test current at increasing values until the RCD trips and then displays the actual current which caused the trip.

As an RCD should not operate at a current of 50 per cent $I_{\Delta n}$ and should always operate at a current of 100 per cent $I_{\Delta n}$ we can use the actual trip current to determine if the RCD is too sensitive.

Example: A 30 mA RCD is reported as tripping repeatedly and for no obvious reason. Having disconnected the load a ramp test is applied to the RCD. If the actual trip current is in the region of 15 mA then the RCD may be over sensitive and could need to be replaced. If the trip current is in the region of 25 to 30 mA then the problem does not lie with the RCD sensitivity and there must be another cause.

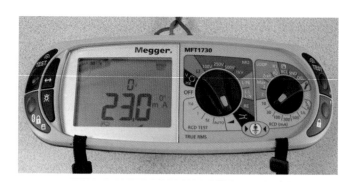

Figure 4.31 *Ramp testing setting*

Phase sequencing

Problems sometimes occur through incorrect phase sequencing of circuits and equipment, the most notable is the reverse direction of three

Try this

1   A 300 mA RCD is to be tested to confirm correct operation. State the maximum test current to be applied and the maximum disconnection time at that test current.

_____

_____

2   Two RCDs are connected in series within an installation. State which type of BS EN 61009 RCD should be installed upstream.

_____

_____

phase induction motors. As a result, a test may need to be carried out to confirm the phase sequencing is correct throughout the installation or circuit.

When an installation is constructed part of the initial verification requires this test to be carried out at:

- The origin of the installation
- Every three phase DB and isolator
- Items of equipment where applicable.

It is not carried out at three phase motor terminals as changing the phase sequence of the windings is used to regulate the direction in which the motor will turn. The test would be performed at the local isolator and starter.

When fault finding, this check could begin at the appropriate point closest to the identified fault and work back towards the origin for the circuit or installation.

There are two recognized types of test instrument used for this test:

- Rotating disc type
- Indicator lamp type.

The installation or equipment needs to be energized for this test and so care is needed when accessing the terminals and making the connections. The instrument is connected in the correct sequence to the three line conductors and the indicator will show the sequence as either correct or incorrect. Once the correct sequence is established at the origin this should be the same at every point in the installation.

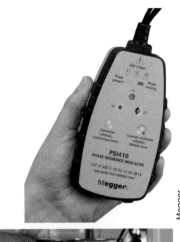

Figure 4.32 *Typical phase sequence tester*

Many instrument manufacturers use colour codes other than those used in the fixed wiring. Also when working on older installations we are likely to encounter the old UK colours. As an aide Figure 4.33 shows a colour comparison for the wiring colours.

If this is the case then the corresponding fixed wiring colours are:

Colour comparison	
New colours	Old colours
Brown	Red
Black	Yellow (or white)
Grey	Blue
Blue	Black

Figure 4.33 *Wiring colours comparison chart*

Considerable care is required when working on older installations as the colours can be misleading, particularly those of blue and black.

- The earlier colour black was a neutral and is now a line
- The earlier colour blue was a line and is now a neutral.

Note

Care is required when connecting and testing and the other line conductor colours (other than black and blue) must be considered to help identify which of the two colour codes is in use.

This phase sequencing throughout the installation may be significant when fault finding to determine the load distribution: where single phase loads are connected to remote DBs the load balancing will have been determined by the designer and incorrect sequence may seriously affect this balance. This may result in one phase of the supply or installation being overloaded whilst others are lightly loaded.

Current and voltage measurement

Measurement of voltage and current is sometimes necessary to determine the existence and cause of faults.

Voltage is measured in parallel to the circuit and this is generally carried out using a suitable voltmeter or the voltage range of a multifunction test instrument. As we discussed earlier, it is always advisable to confirm that voltage is present and determine the level before connecting a voltmeter.

Whilst voltage measurements can be very useful in diagnosing faults on equipment, some caution needs to be used when measuring voltages within an installation. The voltage from the DNO is subject to variations and may change from moment to moment depending on the load on their network. We use a nominal voltage for our calculations and this is taken to be the average voltage likely to be present. The level of tolerance on the DNO supply voltage is between +10% and −6%. For a nominal 230 V supply this is between 253 V and 216 V and whilst fluctuations are not usually across the full 37 V, the changes may be quite significant.

The maximum voltage drop within an installation for a power circuit supplied at 230 V is 5% which would be 11.5 V, so a voltage of 218 V may be acceptable for compliance with BS 7671. This is only true if the supply voltage is at 230 V, but as we have seen, the supply voltage at the origin could be as low as 216 V. In this case, at the extremity of the circuit the voltage could be 204.5 V.

The voltage drop is a product of the resistance of the conductors within the installation and the current flowing in them ($V = I \times R$). The designer will have determined that the cable cross-sectional area (csa) selected will ensure the voltage drop is not exceeded. Measuring voltage at a point on the installation cannot be used to determine the voltage drop within the installation.

Note

More information on voltage drop and determining compliance is given in the *Inspection testing and commissioning* study book in this series.

Current is measured in series with the equipment but this is often not practical and seldom used when fault finding. The most common approach is to measure using a clamp-on ammeter, having jaws which are placed around the line conductor. The jaws form part of a current transformer and the instrument measures the current flowing in the conductor by induction.

Figure 4.34 *Clamp-on ammeter*

Clamp-on ammeters are often multi-ranging but it is important to ensure that the instrument is able to accurately measure the likely current. If in doubt always start on the highest range and work down until a suitable range is selected. If the instrument has ranges of say 10 A, 20 A, 50 A and 200 A then starting at the 200 A range we may have a reading of 40 A. In this case we would reduce the range to the 50 A range to obtain a more accurate reading.

Most clamp-on ammeters also have a voltage measurement scale which requires additional leads (to GS 38) to enable voltage to be measured.

Current measurement is often used to determine possible overloads on circuits and measure leakage currents to earth. If leakage currents are to be measured, the ammeter will need to have a low current range as the leakage values may be in mA.

Figure 4.35 *Low current leakage clamp-on ammeter*

When fault finding on equipment and controls the measurement of voltage and current can be a useful tool. The manufacturer's information generally details the voltages present at points within the control circuits. This information can then be used to identify whether electronic switches are open or closed and whether components are operating correctly. These values may be extra low voltage in the 1 V to 12 V range, and test instruments will need to be able to measure accurately at levels appropriate to the equipment being tested.

Task

Using manufacturer's information or the Internet identify a suitable test instrument and record the manufacturer's catalogue number for each of the following tasks.

1   Confirm phase sequence for a three phase ac motor.

_____

_____

_____

2   Measure the voltage at the terminals of:

   a an extra-low voltage transformer for a boiler control circuit.

_____

_____

_____

   b the terminals of a three phase motor.

_____

_____

_____

3   Measure the current flowing in:

   a a 300 A distribution circuit.

_____

_____

_____

   b the combined functional and protective conductor of an item of IT equipment.

_____

_____

_____

Try this: Crossword

Across

1   This test may be used to identify a faulty RCD (4)

4   The first thing to confirm we have when equipment fails to function (6)

8   Not a ring final circuit (6)

9   May be provided for additional protection (3)

10 How a current measuring meter would be connected (6)

11 A portable source of dc supply (7)

13 An annual event for most test instruments (11)

15 This is not a line or a cpc (7)

16 This is used in 10 above (7)

17 A voltage one occurs over conductor length (4)

18 Ground and where the cpc connects to (5)

Down

2   If the connections are in the right places this is OK (8)

3   Stops us getting on or the ac equivalent of resistance (9)

4   13 A ones are common (6)

5   Installations may be single or three ... (5)

6   Connected in parallel for measurement (9)

7   Needs to be produced during and after fault location and repair (13)

12 Separate from all sources of supply (7)

13 Who pays the bill (6)

14 Not the live conductor coloured blue. (4)

Congratulations, you have finished Chapter 4 of this study book, complete the self assessment questions before continuing to the next chapter.

SELF ASSESSMENT

Circle the correct answers.

1 Which of the following test instruments would be used to identify a short circuit between live conductors and earth?

   a. approved voltage indicator

   b. low resistance ohmmeter

   c. insulation resistance ohmmeter

   d. earth fault loop impedance tester

2 The document used to record that a repaired circuit is safe to put into service is a:

   a. repair record sheet

   b. schedule of test results

   c. electrical installation condition report

   d. minor electrical installation works certificate

3 A circuit is installed in PVC trunking using 1.5 mm^2 copper conductors for the live and cpc. If the resistance of a 1.5 mm^2 conductor is 30.2 mΩ/m and the circuit length is 25 m, the expected measured $R_1 + R_2$ resistance will be:

   a. 15 Ω

   b. 7.5 Ω

   c. 1.51 Ω

   d. 0.76 Ω

4 An insulation resistance test is carried out on a first floor lighting circuit in a dwelling and a tungsten lamp has been left in one luminaire with the control switch open. The effect of this on the insulation resistance test between live conductors will be:

   a. no effect

   b. short circuit

   c. low resistance

   d. approximately 0.01 MΩ

5 A client has recorded that a 30 mA RCD randomly trips out. The most appropriate test to identify whether the RCD is too sensitive is:

   a. operate the integral test button

   b. the electrical function tests

   c. earth fault loop impedance

   d. ramp test

Remedial work

RECAP

Before you start work on this chapter, complete the exercise below to ensure that you remember what you learned earlier.

- The standard voltages which are available on the DNO distribution system are between _____ conductors (U) _____ and between _____ conductors and _____ (U_0) _____ .

- The Health and Safety _____ Guidance Note _____ provides guidance on the requirements for test _____ used by _____ .

- Calibration checks instrument _____ against a _____ standard and confirms the instrument is giving _____ readings.

- Pressure on the electrician to complete fault finding work _____ must _____ lead to taking _____ and risking _____ .

- Work on the _____ wiring of the installation requires _____ to be issued to confirm the _____ of the work with BS 7671.

- Continuity is measured using a Low _____ Ohmmeter and the test leads should be _____ to ensure accurate test _____ .

- The test voltage and minimum acceptable value for an insulation resistance test on a 230 V ac circuit is _____ V and _____ MΩ.

- A polarity test is carried out to confirm all _____ pole _____ and _____ devices are connected in the _____ conductor only.

- The earth fault loop impedance _____ has to be _____ enough to allow the _____ device to _____ within the required _____.

- RCDs are installed to provide _____ protection, earth _____ protection and where there is a _____ of _____.

- The maximum test current to be applied to a 30 mA RCD is _____ mA.

- A ramp test of an RCD applies the test current at _____ values until the RCD _____ and then displays the actual _____ which caused the _____.

- There are two recognized types of phase sequence test instrument, the _____ and _____ lamp type.

- Voltage is measured in _____ with the load, and current is measured in _____ with the load.

- _____ is normally measured using a _____ ammeter.

LEARNING OBJECTIVES

On completion of this chapter you should be able to:

- Identify and explain factors which can affect fault correction, repair or replacement, including:

  - Cost

  - Availability of replacement parts, resources and staff

  - Down time

  - Legal and personal responsibility

  - Access to systems and equipment

  - Provision of emergency or stand-by supplies

  - Client demand.

- Specify the procedures for functional testing and identify tests that can verify fault correction

- State the appropriate documentation that is required for fault correction work and explain how and when it should be completed

- Explain how and why relevant people need to be kept informed during completion of fault correction

- Specify the methods for restoring the condition of building fabric

- State the methods to ensure the safe disposal of any waste and that the work area is left in a safe and clean condition.

Part 1 Factors affecting fault correction, repair or replacement

Careful planning and the sourcing and use of resources will make the fault correction and repair more cost-effective. The first things that have to be determined once the cause of the fault has been located, are:

- Are the necessary replacement parts readily available? This information needs to be established first to allow informed decisions to be made by the client

- Is it viable for the client to have the equipment repaired or is replacement a more cost-effective option? This will need to be discussed with the client and we will need to provide guidance on the suitability of repair, the time and cost options

- When can the work be carried out to minimize disruption and inconvenience to the client?

- Are there appropriately skilled staff available to carry out the necessary remedial work? This will depend on the decisions made with the client regarding repair or replacement and when it is convenient to carry out the work.

Replacement parts

When sourcing replacement parts for items of equipment, the first reference is generally to the manufacturer's information to establish the relevant part number. In some instances we may need to replace one specific item on a printed circuit board. The manufacturer may only supply the circuit board complete and so we may have to source the component from an independent supplier. The information we need may be obtained from the existing faulty component, such as a resistor which has gone open circuit or a failed transistor. It is not always possible to retrieve this information, for example, from a burnt out resistor where the information has been destroyed. In such circumstances we may be able to obtain the information from the manufacturer, many of who have online help and information services.

The considerations here are:

- What was the cause of the failure and will the simple component replacement cure the underlying cause?
- Will the sourcing and replacement of the item involve more time and cost in labour and lost production?
- Would the replacement printed circuit board be more cost-effective and reliable?

We would then communicate this information to the client, enabling them them to make an informed decision.

When repairing or replacing items which form part of the fixed electrical installation, we must ensure that the replacement is suitable for both the job and the environment in which it is to be installed.

Failing to supply suitable equipment or material may result in premature failure or malfunction, which will incur additional cost and disruption to the client. This may also result in action by the client, due to the contractor's failure to provide a suitable service and the additional costs incurred as a result.

For some faults there are no problems in obtaining replacement parts, such as general purpose cables, accessories and the like. There may be a number of availability issues with some of the more specialist items such as luminaires and special finish accessories. These will generally involve some discussion with suppliers and wholesalers to determine the earliest delivery date and best price.

It may prove difficult to negotiate earlier delivery dates and better prices when more specialized equipment is required. This is generally because there is less choice and fewer supplier options available to the contractor.

Figure 5.1 *Seeking the best delivery and price option*

Viability

The viability of repairing faulty equipment should be discussed with the client once the availability and relative costs have been established. The basic formula for determining the viability of repair or replacement includes:

- Overall age and condition of equipment
- Cost of repair vs cost of replacement
- Time for repair vs time for replacement
- Total downtime for each option
- Suitability for the task.

In some instances the age and condition of the equipment alone can be a deciding factor, but generally it is a combination of all of these factors.

Figure 5.2 *Condition can be a deciding factor*

When carrying out fault repairs to electrical installations there are often far less choices available. The installation or circuit is either necessary or not; if it is necessary then a repair is required, if not it can be disconnected and made safe. Of course the installation method for the repair may be different to that of the original , minimizing both disruption and damage to the building fabric. For example, surface mounted enclosures may be used rather than cutting into the building fabric to replace cables concealed in the building fabric.

Downtime

One of the key considerations when repairing faults is the period for which the equipment, circuit or installation will be out of service. We have mentioned in earlier chapters the effects of inconvenience and lost production but there are a number of important factors which need to be determined. Not all of these issues are relevant all of the time, for example:

- The failure of one light in a large, well lit, overground car park may not be a significant issue requiring urgent attention. The time taken to repair the fault and the disruption caused by the work is unlikely to have a significant effect on business.

- Should the circuit supplying the machine used for the first stage of a production process fail, this is likely to be very significant. The whole production process is unable to run without that particular machine and the loss of production, profit, salary and possibly materials can be extremely expensive.

The loss of production can be a major factor in fault finding and we should be conscious of the need to consider the best option for the client. A replacement part which is readily available may be cheaper than a complete component. However, the complete component may mean production will be back up and running more quickly and so the overall cost is lower.

Cost

The main factors which affect the cost of repairs are:

- Spares and materials
- Downtime
- Accessibility
- Equipment
- Labour.

We have considered the spare parts and downtime implications so let's look at the other factors.

Accessibility includes both the timing of the work and the physical access to the work area. If the work has to be done outside of normal working hours this will attract additional cost for the labour involved. This includes the contractor's staff and both the contractor and the clients' staff, as the client will need to provide staff for access and security. The location will also play a part as the ease of access to the worksite for the electrician and the materials will affect the time taken to carry out the work. Where the worksite requires access equipment such as access towers or scaffolding, this will involve additional cost for hire and erection.

Specialized equipment may be required for the remedial work, such as hydraulic crimping tools and conduit bending machines, so the procurement and hire costs will increase the overall cost of the remedial work. Many companies do not keep the more specialized tools and hire them on a needs must basis. General hand and power tools are normal workday commodities and should not affect the costing for the work.

Labour

Labour is a significant factor in cost control which often proves to be a decisive factor in the repair process. At one time it was a common practice for electrical appliances such as irons and kettles to be repaired. Nowadays the cost of the labour to carry out the repair, without the cost of the spare parts, outweighs that of new replacement equipment. If it takes an electrician an hour to replace a fluorescent luminaire and commission it into service, and 2 hours to repair the existing luminaire, the replacement is likely to be the cheaper option.

Once the fault has been located and the remedial action has been agreed, the time it takes to complete a job of work can be measured and predicted with reasonable accuracy. Any interruption or delay to the work activities can cause the work to overrun.

Events such as obstruction of the work, additional congestion, restriction of the work area or its access and disrupted working time due to the activities of others all add time and therefore labour cost to the work. We need to discuss this with the client and ensure that delays and obstructions are kept to a minimum.

Figure 5.3 *Changes to the access arrangements*

Generally the major increase in labour cost is caused by delays to progress. The main causes of these delays include:

- Contract anomalies
- Shortage of resources
- Site constraints
- Environmental conditions
- Third party actions.

Remember

Any shortage or delay in the provision of the plant, material, equipment or operatives to adequately resource the work will result in a delay and additional cost to the contractor.

Particular requirements

Any additional requirements to enable the work to progress will attract additional costs. In addition to those we have already considered there may be legal and contractual issues which will involve additional expense and so additional costs. There may be a requirement for a warranty to be provided for the remedial work and any replacement equipment. Many of these warranties are provided as an insurance backed warranty, which safeguards the client in the event of the contractor reneging on their agreement or going out of business. The provision of these warranties will involve additional cost.

Certain locations will have restrictions on the hours during which noisy activities such as percussion drilling and chasing can be undertaken, due to the working requirements of adjacent tenants. This is particularly common when working on a single floor in an office block where the other floors are still in use.

Inner city projects often have limited access for plant, equipment and operatives. Good planning and liaison can help to reduce the likelihood of delays occurring. This may mean that the operatives will not be able to park vehicles on site and may need to carry equipment and tools some distance. Deliveries may need to be strictly scheduled and some premises may restrict deliveries to outside of normal working hours only.

The problems are not restricted to deliveries. The disposal of rubbish, packaging and redundant plant and equipment will have to be removed from the site. It may not be possible to arrange for a skip to be located on or adjacent to the site. Waste may need to be placed into sacks and stored, with regular pick-up times within the site access schedule.

Such restrictions have a real potential to cause delays to the progress of the work and incur additional cost.

There are occasions where there will be a need to engage the services of specialists to carry out part, or all of the remedial work. For example, the failure of a machine controlled by a programmable logic controller (PLC) may involve the use of a specialist in the operation and programming of the PLC. Similarly, problems with solar photovoltaic supply may require a specialist to determine the problem and remedy.

Temporary supplies

We may need to provide a temporary supply to enable:

- The client to continue some or all of their business activities
- The remedial work to be carried out.

The client may need to have a supply available for some of their business activities to continue during the remedial work.

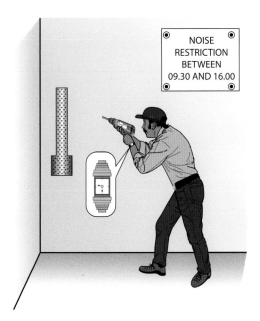

Figure 5.4 *Site constraints may affect productivity*

Simple requirement: this may simply be some temporary lighting and power to allow the office telephone and reception activities to carry on. A simple portable generator may be all that is required to meet this requirement.

Stephill Generators Ltd.

Figure 5.5 *Typical small generator*

Complex requirement: it may involve a considerable temporary supply to enable business functions to carry on very much as normal. For example, a fault on the client's LV switchgear on their 11 kV supply transformer may involve some days or weeks to obtain a suitable replacement. The client cannot afford for business to halt for that period of time and so an alternative supply will need to be arranged. A large transportable generator could be required to allow the work to continue, together with the temporary supply cables, switchgear and connections to make this possible.

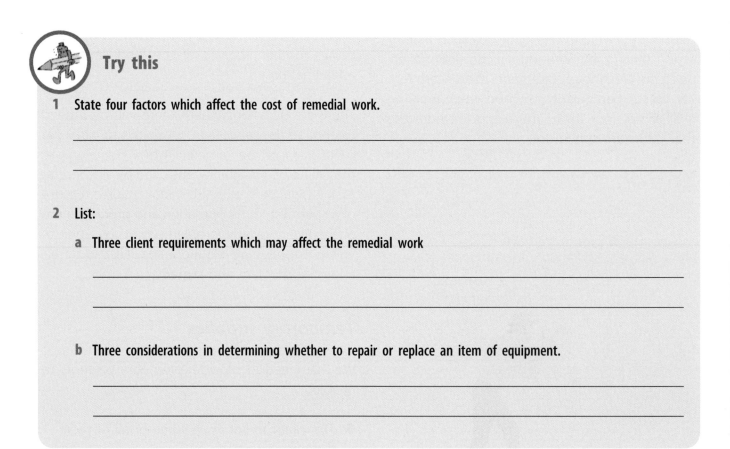

Try this

1   State four factors which affect the cost of remedial work.

_____

_____

2   List:

   a   Three client requirements which may affect the remedial work

   _____

   _____

   b   Three considerations in determining whether to repair or replace an item of equipment.

   _____

   _____

Part 2 Testing to verify fault correction

In the previous chapter of this study book we considered the tests carried out to identify and locate faults. We also have to carry out tests during and on completion of the remedial work. In this chapter we will not repeat the procedures for these tests but identify the tests that are required and the results which would be expected.

Note

More detailed information on all these tests can be found in the *Inspection testing and commissioning* study book in this series.

Continuity

A test will be carried out to confirm the continuity of conductors. At initial verification the tests for continuity relate to:

● Circuit protective conductors
● Protective bonding conductors
● Ring final circuit conductors.

When rectifying faults it is good practice to confirm the continuity of all the conductors during the remedial stages.

Remember

We are measuring conductor resistance so this will involve low values of resistance.

The two basic methods of testing to confirm the continuity of conductors are:

Method 1

This involves linking the line and cpc together at the DB and measuring between the line and cpc at each point on the circuit. This process is best carried out once accessories are connected but before they are fixed into position. Method 1 works well when testing newly installed circuits, conductors and accessories. When testing existing circuits and installations, it is often better to use Method 2 as this requires no interference with the accessories of equipment.

Remember

This test can be carried out using any two conductors when confirming remedial work.

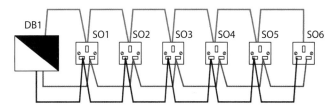

Figure 5.6 *Repaired radial circuit*

Method 2

An alternative method is to use a long or wander lead and test between each point on the circuit to confirm continuity.

Remember

The long test lead should be nulled out before the testing is undertaken.

The length of the test lead may be kept reasonably short by testing from the DB to SO1 and then from SO1 to SO2 and so on.

When confirming the remedial work meets the requirements we can use the test results and confirm using standard values for conductor resistance. IET Guidance Note 3 contains details of conductor resistances in mΩ per metre in Appendix B Table B1.

Using the measured resistance plus the length of the circuit and the standard resistance values we can verify that the conductors are continuous and the value is acceptable.

$$\text{Measured resistance} = \frac{\text{Conductor resistance} \,(\text{m}\Omega) \times \text{Length}}{1000}$$

Ring final circuits: Where the remedial work involves repairs to a ring final circuit, the continuity of the ring final circuit must be confirmed before it is placed into service.

> **Note**
>
> More detailed information on testing continuity of ring final circuits can be found in the *Inspection Testing and Commissioning* study book in this series.

> **Remember**
>
> Where circuits are installed in steel containment systems the steel containment will create parallel paths which will result in the measured cpc values being lower than expected.

Continuity of ring final circuit conductors

Continuity of ring final circuit tests are to be carried out on all ring final circuits at initial verification. This is the first test to be applied to a ring final circuit as once the test is completed the continuity of the protective conductor will have been confirmed.

The purpose of the continuity of ring final circuits is to confirm that:

● The fault has been corrected
● The circuit is a ring
● There are no cross or interconnections in the ring.

Unfortunately it is relatively easy to make a mistake when installing ring circuits, particularly when using single core cables in conduit or trunking. We need to determine that the ring is actually a ring with no cross or interconnections before it is placed back in service.

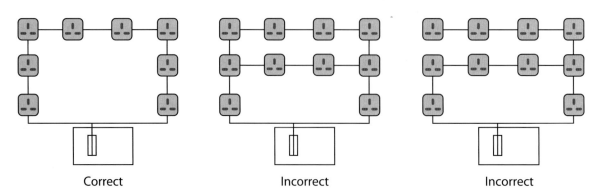

Correct   Incorrect   Incorrect

Figure 5.7 *Line diagrams showing correct and incorrect connections of a ring circuit*

This test is carried out using the low resistance ohm-meter, is carried out in three steps and can be performed at any point on the ring. The test is best carried out from a suitable socket outlet on the circuit, once the ring final circuit is terminated at the DB and the cb locked off.

A proprietary plug-in unit is a useful accessory when carrying out this test as it allows the test to be carried out from the socket fronts. This means the sockets can be fixed during this test.

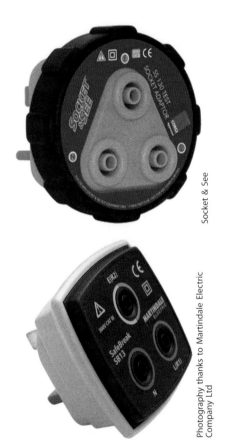

Socket & See

Photography thanks to Martindale Electric Company Ltd

Figure 5.8 *Typical proprietary plug-in unit*

Carrying out the test

Step 1: the first step of the test is to confirm the conductors are connected in a ring and to establish the resistance of each conductor from end to end. The test instrument is connected between the two line conductors, a test is carried out and a reading obtained. This is recorded as r_1.

We repeat the test with the two neutral conductors and the results should be substantially the same, as the cables should follow the same routes. So within 0.05 Ω, this is recorded as r_n.

The test is then repeated with the cpc and the result is recorded as r_2. Where the cpc is a smaller csa than the live conductors the resistance will be proportionally higher.

This value is arrived at by using the proportion of the csa, and as resistance is directly proportional to csa we can establish the ratio by dividing the csa of the live conductor by the csa of the cpc.

Example: $\dfrac{2.5\,\text{mm}^2}{1.5\,\text{mm}^2} = 1.67$ so the cpc resistance should be approximately 1.67 times higher.

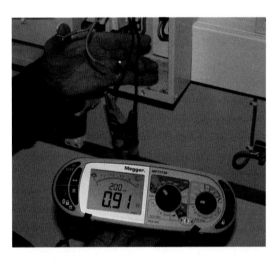

Figure 5.9 *Testing r_1*

If the test result line to line (r_1) on the ring circuit is 0.47 Ω, then we can estimate the total length of the conductor by using the formula length = total resistance ÷ resistance per metre. If the circuit is installed using 2.5 mm² conductors then:

$$\text{Length} = \frac{0.47\,\Omega}{\text{Resistance per metre}} = \frac{0.47\,\Omega}{0.007\,\Omega} = 67\,\text{m}.$$

This is an approximate value which gives us some idea of the total length of conductor involved.

Providing the tests have confirmed that the conductors have been correctly identified and form a ring we can continue to the next step.

Step 2: the next step is to 'cross connect' the line and neutral conductors, that is, connect the line of one end of the ring circuit to the neutral of the other end and vice versa. This is relatively easy when the installation is in sheathed cables but may be a little more difficult when using singles in trunking and conduit.

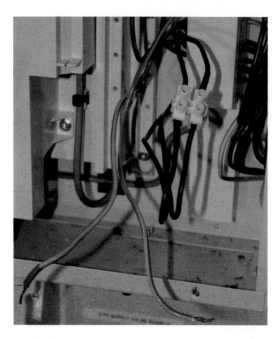

Figure 5.10 *The cross connection of the line and neutral conductors*

The resistance of the conductors is then measured across the connected pairs and the values recorded. These values should be approximately a quarter of the line and neutral resistances added together:

$$\left(\frac{r_1 + r_n}{4}\right).$$

Using a plug adapter, test at each socket outlet between the line and neutral conductors with the cross connections in place.

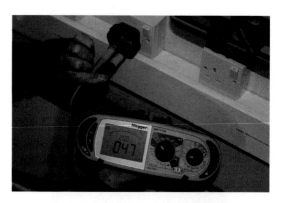

Figure 5.11 *Testing between L and N at the socket outlets*

The test readings obtained at each socket should be substantially the same as the reading taken from the test at the point where the cross connections were made. Socket outlets connected via a spur from the ring circuit will give higher values of resistance, and the precise increase will be proportional to the length of the branch cable.

Step 3: repeat the process from Step 2 only this time we cross connect the line and cpc of the ring circuit. Where the cpc is a smaller csa than the live conductors the resistance will be higher than in the previous test. Again test at each outlet, between line and cpc this time. Again the readings obtained at sockets connected via a spur will be higher.

a)

Courtesy of Kewtech

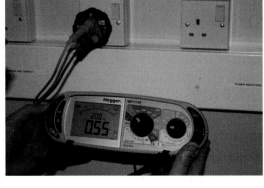

b)

Figure 5.12 *Test between L and cpc at every socket*

The highest value obtained during this step of the test represents the $R_1 + R_2$ value for the ring circuit.

The value obtained should be in the order of:

$$\left(\frac{r_1 + r_2}{4}\right)$$

using the $r_1 + r_2$ values obtained in Step 1.

If we have carried out Steps 2 and 3 from the socket fronts using a socket adapter, the polarity of the circuit will also have been confirmed.

Try this

The continuity of a ring final circuit is being tested and at Step 1 the results are $r_1 = 0.30\ \Omega$, $r_n = 0.31\ \Omega$ and $r_2 = 0.52\ \Omega$. All the socket outlets are connected into the ring.

Determine:

a the expected value at each socket when the line and neutral are correctly interconnected.

_____

_____

b the expected value at each socket when the line and cpc are correctly interconnected.

_____

_____

Insulation resistance

Insulation resistance testing is used to confirm the resistance of insulation between live conductors and live conductors and Earth is acceptable.

Nominal circuit voltage	Test voltage (dc)	Minimum insulation resistance
SELV and PELV only	250 V	0.5 MΩ
Up to and including 500 V (excluding SELV and PELV only)	500 V	1.0 MΩ
Over 500 V	1000 V	1.0 MΩ

Figure 5.13 *Insulation resistance test applied voltages and minimum acceptable values*

BS 7671 identifies the minimum acceptable insulation resistance values and the applied test voltages for installations (not circuits).

When we are measuring insulation resistance following remedial work we would expect healthy circuits to have values far higher than the minimum values in BS 7671.

There are two main safety considerations when carrying out an insulation resistance test:

● The circuit must be isolated from the supply
● The test voltages being injected into the installation are high and could cause danger
● Whilst the output is unlikely to result in a fatal electric shock the involuntary reaction to the shock can result in falls from height and other physical injuries.

The test is then carried out on the disconnected circuit between all the live conductors and all live conductors and Earth. BS 7671 and IET Guidance Note 3 identify that when carrying out this test to confirm new work is compliant:

● The protective conductor must be connected to the main earthing terminal
● All the protective bonding must be connected.

For a healthy new circuit in good condition it is quite possible that the reading will indicate the resistance is over the range of the instrument.

Note

When testing on existing installations it may be advisable to perform a test L-N at the lower voltage, 250 V dc, before testing at the required voltage.

When testing three phase and three phase and neutral circuits the test should be carried out between all live conductors and all live conductors and Earth, just as we did for the single phase installation.

Polarity

The polarity of a circuit may have been confirmed using the tests carried out when testing continuity. In any event the polarity must be confirmed before the circuit or installation is energized. Either of the continuity test methods may be used to confirm polarity. Where Method 2 is used the appropriate line conductor should be used.

Remember

Confirming polarity is carried out to ensure that:

● **All single pole protective and control devices are connected in the line conductors only**
● **The centre pin of ES lampholders are connected to the line conductor only**
● **All equipment and socket outlets are correctly connected.**

Earth fault loop impedance testing

Following the remedial work and the tests which must be carried out before the circuit is energized we need to confirm the earth fault loop impedance is acceptable.

System earth fault loop impedance (Z_s)

The test instrument used is the earth fault loop impedance tester. The test is to confirm that the earth fault loop impedance Z_s is low enough to allow the protective device to disconnect within the required time.

In the event of a fault to earth the fault path consists of the circuit line conductor and cpc ($R_1 + R_2$) and the external earth fault loop path (Z_e), so we can determine the expected Z_s value for a circuit using the formula $Z_s = Z_e + (R_1 + R_2)$.

Remember

The earth fault loop impedance test is carried out with the conductors at a temperature lower than their normal operating temperature.

Where it is not possible to directly measure earth fault loop impedance, the value of Z_s can be determined by calculation using the formula: $Z_s = Z_e + (R_1 + R_2)$.

Remember

The measured values of both Z_e and $R_1 + R_2$ must be used when determining Z_s.

Once the value of Z_s has been obtained this must be compared to the maximum acceptable value for the protective device fitted for the circuit.

This is done by either:

- Applying the 0.8 multiplier to the maximum tabulated values in BS 7671
- Using the temperature corrected tables contained in IET Guidance Note 3 or the IET On-Site Guide.

The measured value must be ≤ the temperature corrected value for the protective device.

If the system of supply is TT then, in all but very exceptional circumstances, an RCD will be required to provide earth fault protection. The value of earth fault loop impedance for these circuits depends upon the rating of the RCD used.

Maximum external earth fault loop impedance values for TT systems	
RCD $I_{\Delta n}$	Maximum R_A in Ω
30 mA	1667 Ω
100 mA	500 Ω
300 mA	167 Ω
500 mA	100 Ω

Figure 5.14 *Maximum earth fault loop impedance values for RCD protected circuits*

Testing residual current devices

> **Note**
>
> Before testing RCDs we must confirm there is a suitable earth fault path and so an earth fault loop impedance test must be carried out first. This may be done from the incoming terminals of the RCD.

An RCD test instrument is used for the test and the RCD should respond within the maximum operating parameters shown in Figure 5.15. The test is carried out in both the positive and negative half cycles of the supply and the highest value obtained is recorded in each case.

RCD operating times		
Test current	Operating time	Time delay or S type
$5 \times I_{\Delta n}$	40 ms	
$1 \times I_{\Delta n}$ (BS 4293)	200 ms	200 ms + 50 to 100% of delay
$1 \times I_{\Delta n}$ (BS EN 61008 & 9)	300 ms	130–500 ms (S type)
$\frac{1}{2} \times I_{\Delta n}$	No trip	

Figure 5.15 *Maximum operating times for RCDs*

> **Note**
>
> The five times $I_{\Delta n}$ test is only carried out on RCDs $\leq$ 30 mA which are used for additional protection.

Phase sequencing

Where work has been carried out on three phase circuits a test needs to be carried out to confirm the phase sequencing is correct throughout the installation or circuit.

The test is carried out at the relevant points in the installation or circuit which may include:

- The origin of the installation
- Every three phase distribution board and isolator
- Items of equipment where applicable.

The types of test instrument which are used for this test are:

- Rotating disc type
- Indicator lamp type.

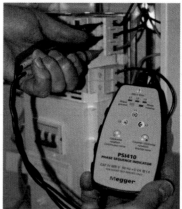

Figure 5.16 *Typical phase sequence tester*

Current and voltage measurement

Measurement of voltage and current may be necessary following remedial work. This is more common when repairs have been carried out on items of equipment. It is occasionally necessary to confirm the voltage and current for a circuit or installation once remedial work has been completed.

Voltage is measured in parallel to the circuit and is generally carried out using a suitable voltmeter or the voltage range of a multifunction test instrument.

Current is measured using a clamp-on ammeter having jaws which are placed around the line conductor.

Figure 5.17 *Clamp-on ammeter*

 Try this

1   A 16 A BS EN 60898, Type C circuit breaker protects a power circuit. The maximum tabulated value of earth fault loop for this circuit is 1.44 Ω. The measured earth fault loop impedance at the equipment supplied on this circuit is 1.2 Ω. Determine whether the circuit meets the requirements for fault protection.

_____

_____

2   A 30 mA BS EN 61009 RCD is to be tested to confirm correct operation. State the maximum test current to be applied and the maximum disconnection time at that test current.

_____

_____

Part 3 Certifying remedial work

Earlier in this study book we considered the documentation required during the investigation and rectification of faults. This part serves as a reminder of the documentation which is to be completed following the remedial work. This documentation is used to demonstrate the compliance of the remedial work and that the circuit, installation or equipment is safe to be placed in service.

Certification of remedial work

When a fault on the electrical installation has been rectified the work must be certificated to confirm the installation is safe for use and complies with the requirements of BS 7671.

For work on a single circuit this will normally involve using a Minor Electrical Installation Works Certificate (MEIWC).

Figure 5.18 *MEIWC for the certification of remedial work*

The MEIWC produced by the Electrical Contractors Association (ECA) allows for up to three circuits to be certified using the MEIWC. However, most fault repairs usually involve a single circuit.

Where more circuits are involved then either a number of MEIWCs must be issued or an Electrical Installation Certificate (EIC) may be used. This EIC must be accompanied by the appropriate

Schedule of Inspections and Schedule of Test Results in order for the certificate to be considered valid.

The limitations on the issue of a MEIWC are such that where work like the replacement of a DB or a consumer unit is carried out, an EIC must be issued. There are certain essential tests which should be carried out on all the circuits reconnected to the DB and these include:

- Continuity of protective conductors
- Earth fault loop impedance
- Polarity.

Figure 5.19 *Electrical Installation Certificate and the Schedules*

Where faults involve equipment and machines the recording process is slightly different.

Some companies use their maintenance records whilst others have a separate document for recording the fault and remedial action. Smaller companies and private individuals often do not have any bespoke forms for this purpose and so many contractors produce their own.

An example was given in Chapter 1 of this study book and is repeated in Figure 5.20, on page 117, as a reminder of the details recorded.

Charles machinists Ltd		Repair record				No. 00173
Date	Equipment Ref No	Fault	Symptoms	Cause	Material	Re-commissioned
--/07/20--	Workshop heater WH1	No heat and fan boost not operating	Operating OK and then stopped working	Traced to thermal cut-out fused	Replacement thermal cut-out TK4-R120	--/07/20—am.

Name: James Douglas   Signature: *James Douglas*   Date: --/07/20--

Figure 5.20 *Typical repair record sheet*

Task

1 A repair to a client's car park lighting circuit involved the replacement of a cable supplying two columns at the end of the cable run. State the form of certification to be issued to the client on completion of the work.

_____

_____

2 Following a major fault the distribution board for one floor of a multi-storey office building has been replaced. List all the documents that are to be completed by the contractor and issued to the client once the work is completed.

_____

_____

Part 4 Keeping others informed

During the fault rectification remedial work we need to keep others informed of the progress. This includes:

- The client
- Colleagues
- Staff
- Other trades.

The progress of sourcing materials or parts and of the repair process will be particularly important to the client. Other activities need to be advised appropriately as the work progresses.

The client needs to be kept informed of the progress of the work and given any information which may affect the date when the work will be completed. Any problems with access or disruption to the work will need to be discussed urgently to minimize the effects.

Figure 5.21 *Keeping the client up-to-date*

In addition to keeping the client informed we need to keep work colleagues advised on the progress of the work. Work on the repair may involve more than one person and each operative needs to advise the other of their progress. Coordination between operatives is important when carrying out any work on site and fault finding is no exception. The progress of the work will depend upon each person completing their tasks on time. Delay to one activity can have a knock-on effect and compound the time taken as well as the cost of the job.

Figure 5.22 *Keeping colleagues updated*

The staff and operatives affected by the work area should be kept aware of your work activities. Their cooperation and understanding plays an important part in the smooth running of the work. There is bound to be some disruption but it is important that people understand what is being done and have some indication of the time it will take. The sooner we complete our work the sooner things are back to normal and the cooperation of the staff will help speed things up.

The remedial work may also involve other trades, including specialist trades, and we need to keep everyone informed on the progress. Many of the processes will be interrelated and dependent upon other works being completed. The supply may be necessary for the specialist contractor to set up and commission equipment so they will need to know when that will be. Equally, equipment will need to be isolated for mechanical engineers to work on it.

Often the repair work is of short duration and the longest period is in obtaining the spare parts. In many straightforward cases the whole process may be completed in a day or two.

The communication during these short work period jobs is generally carried out verbally and only where there is a considerable change is written communication required. Where no other contractors are involved the client may be kept informed by a brief meeting at the end of the day to update them on progress. Any changes such as the use of different equipment or materials will need to be confirmed in writing. Any changes to the agreed working pattern, such as change of access or availability of an area will need to be discussed and formally agreed. These changes would be formally agreed using a variation order.

More involved fault finding and repair may take some time and once the fault has been located, the programming of the remedial work can be undertaken. Like any other project the attendance of various trades and specialists on site will need to be programmed and standard project planning processes would normally be used.

Note

More details on the planning and administration of the work can be found in the *Organizing and Managing the Work Environment* study book in this series.

Once the rectification work has begun, there should be regular site meetings with the other trades to discuss progress and identify any events that will affect the progress of the work. This may be due to an unexpected problem and so introduce a delay, or work may have progressed better than expected and activities can be brought forward.

Once a contract has been agreed, any changes to the original work will need to be confirmed using a variation order. Those that apply to our activities will need to be issued by the client, or their representative. Any which apply to the activities of our sub-contractors or specialist trades would then be issued by your company.

The regular meeting and progress reports are an important part of the work process and should not be underestimated. These meetings may be quite short where all is going to programme, but allow any problems or gains to be identified at an early stage. This enables all parties to take the necessary action to recover the programme or achieve early completion.

VARIATION ORDER

Issued by:	George Edward Associates
Employer:	Rachel Louise Hairdressing
Contractor:	Arual Development Ltd
Works situated at:	Rachel Louise Hairdressing, Linking Road
Job Ref.:	C535
Issue Date:	01-07-9-

Under the terms of the above contract, I/We issue the following instructions:

No.	Instruction	Office use: £ omit	Office use: £ add

Signed:

Amount of Contract Sum	£
Approx. Value of previous instructions	£
Approx. Value of this instruction	£
Approx. adjusted total	£

Figure 5.23 *Typical variation order*

The sense of urgency which often accompanies fault finding and remedial work should not be allowed to completely bypass the contractual requirements for the work. For expediency, many contractors use a site instruction form which allows the person responsible for the site to issue and receive instructions. These site instructions are then followed up by a variation order to confirm the requirements. This process allows the work to progress with minimum disruption or delay whilst waiting for paperwork.

SITE INSTRUCTION FORM

SITE INSTRUCTION			
CONTRACT NO.			
FROM		TO	
SITE INSTRUCTION NO.		DATE	
FILE REF.		COPIES	

Attachments: _____

_____
Client's Representative

Received by :

_____
Contractor's Representative

Date : _____   Time: _____

Figure 5.24 *Typical site instruction form*

Try this

Remedial work to a vacuum moulding machine is being carried out to replace the injection nozzles. The client has requested that the machine is serviced whilst this repair is being carried out. Using the form in Figure 5.24, complete a site instruction to the specialist contractor requesting that a full service of the injection moulding machine is carried out.

Part 5 Making good and clearing up

When fault finding and repair is undertaken, the building should be reinstated to a condition equivalent to its state when we arrived. There is often some damage to the building fabric due to the need to access cables and accessories. In some instances obtaining access may be carried out without causing any damage and so there will be no need for making good. The removal and replacement of ceiling tiles to access lighting circuits is a good example of this.

In other cases there may be a need to plaster up chases and some painting to restore the worksite to the original standards of decoration. In extreme cases there may be major building works to be undertaken involving building and finishing contractors.

We shall look at the basic requirements for repairs to the building structure that would reasonably be within our scope of work.

Where cables and containment systems pass through the building structure the reinstatement must have the same properties as the original structure. That means the same physical strength, density, fire and sound resistance. Whilst it is not always possible to use an identical material a suitable alternative must be used.

Some reinstatement may be simply cosmetic, others may be structural and some have elements of both. Holes left unfilled can lead not only to the spread of fire and/or noise but also allow access to rodents and insects.

Reinstatement of chases cut into plaster for cables are generally easily repaired using plaster mixed on site. This should be applied to dampened brickwork and surrounding plaster. All loose material should be removed and the surrounding material wetted to prevent the plaster drying out too fast and not bonding correctly. It may be useful to use a PVA bonding agent when doing this as it helps with the adhesion of the new plaster.

Figure 5.25 *PVA bonding agent*

It may also be necessary to apply the plaster in more than one coat depending on the extent and depth of the plaster. It is not uncommon for the first coat of plaster to crack, particularly where the cable or capping etc. is installed in the chase. A two-coat approach generally overcomes the problem and whilst the final finish may still show some cracks these can normally be filled using proprietary filler.

Where containment systems pass through the building structure the surrounding fabric needs to be reinstated. If a trunking is installed then a section of lid will need to be installed through the building fabric. It is good practice for this not to extend too far into the building, but just through the structure, with the removable lid running up to this section of fixed lid.

When reinstating the fabric of the building, the surrounding gaps should not be stuffed with paper and then cement applied to the outer surface. This will not maintain the structural or fire integrity of the building. The reinstatement material should fill the building material void completely and provide equal mechanical strength and fire resistance.

The use of a plasticizer in the mortar mix for these repairs makes the mortar easier to work with and provides better adhesion. As with the plaster, all loose material should be removed and the surrounding material wetted to prevent the mortar drying out and not bonding correctly.

Bostick

Figure 5.26 *Mortar plasticizer*

Where cables pass through ceilings, for example in a domestic property for lights, luminaires and the like, there is inevitably going to be a hole around the cable. Filling these can be difficult using plaster or filler and movement of the cable can result in the filler coming away.

Most ceilings in these situations provide at least a half hour fire protection, so any holes must be suitably filled to maintain this. Proprietary fire resistant mastic is available which can be installed using a mastic gun. This material has stated levels of fire protection and generally does not set hard, allowing for some movement due to changes in cable temperature etc. This makes the filling of the holes and compliance with the requirements of BS 7671 and the Building Regulations relatively easy.

Figure 5.27 *Fire sealant mastic*

Where cables or containment systems pass through floors, the fire stopping material must be secure to the surrounding material and able to withstand the fire conditions likely to be encountered. It must also have the same expansion characteristics as the material, because the fire stopping should not fall out in the event of a fire because of different expansion rates.

Figure 5.28 *Fire stopping*

When floorboards and chipboard flooring needs to be refitted following electrical installation work, it is important to ensure that the cables installed are not damaged in the process, or as a result of moving the boards.

Where floorboards are refitted, this may be done using nails through the board into the joists which should not be in the same location as the original nails. This is because nails rely on the displacement of the wood fibres for their grip. Once the nails are removed, the grip reduces and the floorboards work loose and start to creak. One way of preventing this is to fix the floorboards using woodscrews in place of nails.

Where cables pass through the building fabric, either on a cable tray or independently supported, a proprietary fire stopping may be installed. This allows the fire stopping material to be both fitted to the cables and, in most cases, removable for alteration or additions to the installation.

Where replacement and tracing of cables has been undertaken it is quite possible that the fire stopping within the containment system has been removed or damaged.

Chipboard floors present similar problems and often the original floor would have been laid using ring nails for better fixing. These must not be replaced in the same location for the same reasons as floorboard nails. Again the option of screwing the boards down may prove more advantageous.

Once the reinstatement is complete the appearance should match and blend with the original finish. Where specialist materials, finishes or complex remedial work is required it is advisable to engage the services of a suitable professional to undertake the work on your behalf.

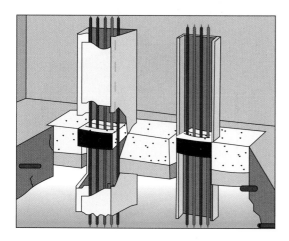

Figure 5.29 *Fire stopping materials in enclosures*

Figure 5.30 *Specialist materials may require specialist removal and replacement*

Legislation holds companies responsible for the waste produced during the course of their work right up to the point of disposal. This means that an appropriate provision for waste disposal must be set up on site.

It may be that the client has suitable waste disposal facilities already on site and may give permission for these to be used during the fault finding and repair process. Of course there may be materials that cannot be disposed of using the client's facilities and so we must make suitable provision ourselves.

The waste which is likely to be produced will be either recyclable or non-recyclable and fall into one of the following categories:

● Waste that is not considered to be hazardous
● Waste considered to be hazardous under the hazardous waste regulations
● Waste that needs to be assessed to find out whether it is hazardous or not.

A designated point on site should be agreed with the client where any waste material is to be stored awaiting disposal. During the working day smaller containers may be used to collect waste material which will then be transported to the storage area.

Within the storage area there should be suitable storage facilities for each type of waste that is produced. This waste is then to be collected by an authorized waste disposal contractor, licensed for the specific materials being disposed of, or recycled.

A number of waste products may be produced during the fault finding and repair process. Depending on the nature of the work there may be relatively small quantities which the contractor can readily remove from site. On more involved projects there may be large quantities of waste material some of which may require specialist disposal.

© Image reproduced with permission of HIPPOWASTE

Figure 5.31 *Typical waste disposal facilities*

Some of these common waste products and their method of disposal include:

- *Paper and cardboard*: this is widely recyclable and should be sent to the local recycle centre.
- *Cable and cable off cuts*: these are recyclable by specialist contractors separating the components and recycling all the recyclable products.
- *Building rubble*: this may be recycled or sent to landfill sites. It is generally sent off site in skips or grab/bulk bags. Large quantities of building rubble are often collected using a grab lorry and loaded direct from the site location.
- *Electrical fittings and accessories*: these need to be disposed of under the WEEE directive and much of their content may be recyclable but will need special facilities to do so.
- *Discharge lamps*: these are recyclable and require specialist treatment. Approved disposal contractors will provide suitable on-site

storage facilities if the quantities are large enough. Otherwise local recycle centres generally have a facility for these lamps. They should **not** be broken up on site.

In smaller properties such as dwellings it may be necessary for us to dispose of the waste produced by our work activities. This waste should still be segregated and taken to a suitable recycling and disposal centre. These are often run by, or on behalf of, the local authority. These have specific areas for hazardous and recyclable waste and an area for safe general 'landfill' waste.

These facilities make a charge for recycling commercial waste (generated by companies), but waste from the general public within the local authority area is usually disposed of free of charge.

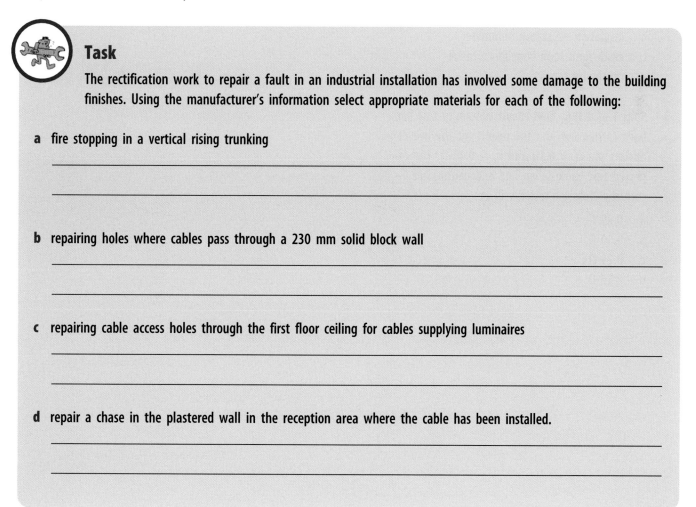

Task

The rectification work to repair a fault in an industrial installation has involved some damage to the building finishes. Using the manufacturer's information select appropriate materials for each of the following:

a fire stopping in a vertical rising trunking

_____

_____

b repairing holes where cables pass through a 230 mm solid block wall

_____

_____

c repairing cable access holes through the first floor ceiling for cables supplying luminaires

_____

_____

d repair a chase in the plastered wall in the reception area where the cable has been installed.

_____

_____

Congratulations, you have now finished this study book. Complete the self assessment questions before going on to the end test.

SELF ASSESSMENT

Circle the correct answers.

1   Which of the following is not a factor which will affect the remedial work carried out on a fault:

a.   availability of spares

b.   working out of hours

c.   access to the worksite

d.   the equipment supplier

2   The presence of a cpc at each accessory following the replacement of a cable is carried out using a:

a.   low resistance ohmmeter

b.   insulation resistance ohmmeter

c.   earth fault loop impedance tester

d.   continuity tester

3   Step 1 of a ring final circuit continuity test has been carried out and the results for the live conductors are $r_1 = 0.5\,\Omega$ and $r_n = 0.49\ \Omega$. The expected test value when line and neutral are correctly cross-connected will be:

a. $0.99\ \Omega$

b. $0.50\ \Omega$

c. $0.24\ \Omega$

d. $0.12\ \Omega$

4   A Minor Electrical Installation Work Certificate can be used to certify:

i   The replacement of a cable forming part of a circuit

ii   The replacement of a distribution board

Of the above statements:

a.   both statements are correct

b.   both statements are incorrect

c.   only statement i. is correct

d.   only statement ii. is correct

5   To enable work to progress with minimum delay changes to the requirements may be instructed using a:

a.   purchase request

b.   site instruction

c.   revised contract

d.   job sheet

End test

Tick the correct answer

1. A fault has been reported on a metal guillotine. The electrician should ask the client for the maintenance schedule for the guillotine and the:

☐ a. Manufacturer's warranty

☐ b. Manufacturer's instructions

☐ c. Electrical Installation Condition Report

☐ d. Minor Electrical Installation Works Certificate

2. The most suitable document to issue to the client following the replacement of a faulty socket outlet is a:

☐ a. Minor Electrical Installation Works Certificate

☐ b. Electrical Installation Condition Report

☐ c. Electrical Installation Certificate

☐ d. Maintenance schedule

3. Which of the following will NOT affect time taken to complete the fault finding and repair process?

☐ a. The availability of manufacturer's information

☐ b. The experience of the electrician

☐ c. The type of test instrument used

☐ d. The availability of spare parts

4. Communication with the client regarding fault diagnosis should always be clear, courteous and:

☐ a. Verbal

☐ b. Written

☐ c. Technical

☐ d. Accurate

5. During the discussions with the client regarding the work program for locating a fault, one of the key areas to be considered is the need for:

☐ a. Replacement parts

☐ b. Isolation of circuits

☐ c. Total cost of the work

☐ d. Materials required

6. The information that would be required to identify a fault on a processing machine includes the:

☐ a. Electrical Installation Certificate

☐ b. Manufacturer's information

☐ c. Manufacturer's warranty

☐ d. Maintenance records

7. **Ventricular fibrillation may occur when a current flow across the chest of a person is in the region of:**

☐ a. 50 mA

☐ b. 40 mA

☐ c. 30 mA

☐ d. 25 mA

8. **When confirming safe isolation of a three phase distribution board the total number of tests to be made is:**

☐ a. 3

☐ b. 4

☐ c. 7

☐ d. 10

9. **Test instrument leads and probes must be compliant with:**

☐ a. BS 7671

☐ b. GS 38

☐ c. GN3

☐ d. OSG

10. **A fault has been identified on the local isolator for one of the six machines supplied on the same circuit. The effect of the actions necessary to correct this on the other machines will be that the:**

☐ a. Machines will not function

☐ b. Machines will function normally

☐ c. Use of the machines will be restricted

☐ d. Performance of the machines will be limited

11. **The requirements for a competent person carrying out fault finding are that they are familiar with the test instruments to be used and:**

☐ a. Are supervised by a competent person

☐ b. Are careful when working on existing circuits

☐ c. Only work on live equipment when told to do so

☐ d. Are familiar with the equipment being worked on

12. **Having confirmed a fault exists and collected all the available information, the next step would be to:**

☐ a. Order spare parts

☐ b. Isolate the circuit

☐ c. Isolate the installation

☐ d. Analyze the information

13. **A packing bailer has seized up and stopped working. Other equipment supplied on the same circuit has also stopped working. The likely cause of the other equipment not working is:**

☐ a. The supply to the installation has been disconnected

☐ b. The local control fuse for the bailer has operated

☐ c. The circuit protective device has operated

☐ d. A short circuit on the circuit

14. When investigating a loss of supply to an item of equipment the tests would begin at the equipment itself and then work:

☐ a. Upstream

☐ b. Downstream

☐ c. From the origin

☐ d. Randomly around the circuit

15. The most likely cause for the contact condition shown in Figure 1 below is:

☐ a. Short circuit current

☐ b. Earth fault current

☐ c. Overloading

☐ d. Arcing

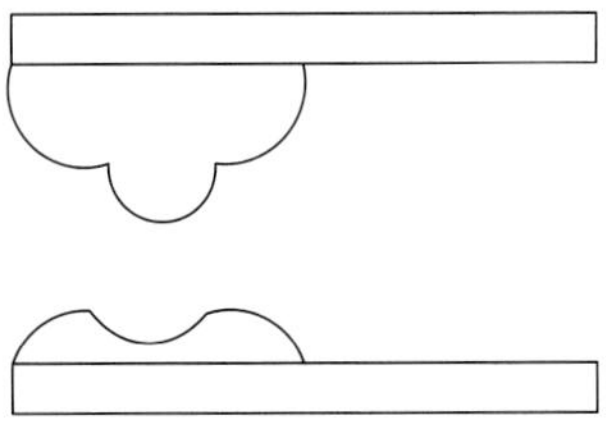

Figure 1

16. A fluorescent luminaire does not light but the tubes flash repeatedly. The most likely cause is:

☐ a. Failed lamp

☐ b. Overvoltage

☐ c. Faulty starter

☐ d. Reduced voltage

17. The luminaires in an open plan office operate but only remain on for 20 minutes before the circuit breaker trips. Investigation reveals that the luminaires have recently been upgraded. The most likely type of fault is:

☐ a. Reduced voltage

☐ b. An overload of the circuit

☐ c. An earth fault on the circuit

☐ d. A fault between live conductors

18. An open circuit fault on a ring final circuit is to be located and the most suitable test instrument for this is:

☐ a. An approved voltage tester

☐ b. Low resistance ohmmeter

☐ c. Insulation resistance tester

☐ d. Earth fault loop impedance tester

19. A short circuit fault between neutral and cpc has been identified on a radial circuit. The most expedient first step in locating the fault is to:

☐ a. Remove all the accessories and test at each point

☐ b. Disconnect the circuit at the mid point and test each half

☐ c. Measure the fault resistance and calculate the length from the origin

☐ d. Test from each end of the circuit and subtract the lowest value from the highest value

20. A suspected low voltage fault at a packing machine has been reported. Before using a voltmeter to measure the voltage the first action would be to:

☐ a. Check the voltage with an approved voltage indicator

☐ b. Calculate the maximum permitted voltage drop

☐ c. Carry out an earth fault loop impedance test

☐ d. Safely isolate the circuit from the supply

21. A circuit breaker protecting a socket outlet circuit in a workshop operates after the workshop has been operating for approximately 1 hour. To determine whether the circuit is being overloaded a test would be carried out using a:

☐ a. Low resistance ohmmeter

☐ b. Clamp meter

☐ c. Wattmeter

☐ d. Voltmeter

22. An insulation resistance test is carried out on a first floor lighting circuit in a dwelling and a tungsten lamp has been left in one luminaire with the control switch closed. The effect of this on the insulation resistance test between live conductors will be:

☐ a. No effect

☐ b. Short circuit

☐ c. Low resistance

☐ d. Approximately 0.01 MΩ

23. A length of cable in a radial socket outlet circuit has been replaced following a fault. The most suitable document to be completed and given to the client is a:

☐ a. Minor Electrical Installation Works Certificate

☐ b. Electrical Installation Condition Report

☐ c. Schedule of Test Results

☐ d. Repair Record Sheet

24. The availability of spare parts will have an effect on the:

☐ a. Time taken to rectify a fault

☐ b. Need to work out of hours

☐ c. Method statement

☐ d. Cost of the repair

25. Following the installation of three replacement luminaires on a workshop lighting circuit the first test to be carried out is:

☐ a. Insulation resistance

☐ b. Earth fault loop impedance

☐ c. Continuity of cpc

☐ d. Polarity

26. Step 1 of a ring final circuit continuity test produced the following test results: $r_1 = 0.37\ \Omega$, $r_n = 0.38\ \Omega$ and $r_2 = 0.62\ \Omega$. The expected result when the line and cpc are correctly cross-connected is:

☐ a. 0.18 Ω

☐ b. 0.25 Ω

☐ c. 0.46 Ω

☐ d. 0.99 Ω

27. A Minor Electrical Installation Works Certificate may be used to certify:
    i   The installation of additional socket outlets to a circuit
    ii The replacement of a faulty socket outlet
    Of the above statements:

    ☐ a. Both statements are correct

    ☐ b. Both statements are incorrect

    ☐ c. Only statement i) is correct

    ☐ d. Only statement ii) is correct

28. To confirm a protective device will operate within the required time in the event of a fault to earth, the test to be undertaken is:

    ☐ a. Continuity of cpc

    ☐ b. Earth fault loop impedance

    ☐ c. Insulation resistance

    ☐ d. Prospective fault current

29. Using a site instruction will allow fault finding work to be progressed without:

    ☐ a. Method statements

    ☐ b. Stage payments

    ☐ c. Supervision

    ☐ d. Delay

30. The replacement of ceiling roses in a dwelling following water damage has resulted in cable entry holes in the ground floor ceiling. The holes must be repaired with a filler which is:

    ☐ a. Colour matched

    ☐ b. Waterproof

    ☐ c. Fire rated

    ☐ d. Rigid

Answer section

Chapter 1

Task Page 7

- The type of light fittings and lamps?
- When did the problem first appear?
- Did they stop working one at a time or did all the other lights go at once?
- How are the lights controlled?
- Is there any obvious damage to the lights or their supports?
- Does the client have spare lamps?

Task Page 9

1 EIC, MEIWC, EICR or PIR
2 MEIWC

Try this: Crossword Page 10

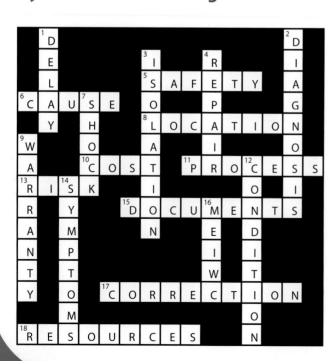

SELF ASSESSMENT Page 11

1 c. maintenance schedule
2 b. isolation of equipment
3 a. accurate
4 d. lighting circuit
5 b. take longer

Chapter 2

Recap Page 12

- initial corrected testing
- logical effective
- resources minimizing isolated
- circuit Installation Certificate
- rectify danger delay
- users informed actions diagnosis repair
- layman's understood

Task Page 17

a The cardiac muscles of the ventricles
b External cardiac compression
c Maintaining blood circulation and oxygen to the vital organs

Task Page 22

1 Permission to isolate the equipment and relevant circuit where necessary (as single machine)
2 a. Informing people in the location where the work is to be undertaken
   b. Use of barriers and warning signs around the work area
   c. Safe isolation

Task Page 28

L1 to N   L2 to N   L3 to N   L1 to E   L2 to E
L3 to E   N to E   L1 to L2   L1 to L3   L2 to L3

SELF ASSESSMENT Page 32

1 b. equipment will function normally
2 c. cannot be inadvertently energized
3 a. confirm the accessory is isolated
4 d. 500 mA
5 a. the test instruments to be used

Chapter 3

Recap Page 34

- competent undertake safety others
- unreasonable dead reasonable on near suitable prevent
- sure switched cannot inadvertently
- burns entry within path
- determine appropriate equipment
- permit control authorize
- cordon safe unauthorized entering
- Warning entry access warn electrical
- tests live live earth Approved tested functioning proving

Task Page 39

Answer: The process should include at least

Task Page 44

Poor termination/worn socket tubes

Dimming circuit failed

Loose lamp/worn centre pin connection

Failed capacitor on lead circuit

Switched to timer function

Task Page 48

Damage to cable from fixings

Circuit overloaded

Operation of item of equipment such as freezer

Open circuit on the lighting circuit at or after the sixth light

Task Page 51

There are a number of answers to these questions and so only a sample is given here.

a Open circuit due to poor termination and short circuit due to rodent damage
b RCD fails and disconnects supply and RCD operated by fault on an item of equipment
c Incorrect connection of strappers and common and burnt out contacts
d Failed element giving low heat and thermal cut out operating

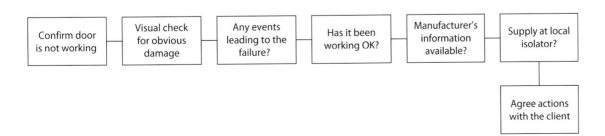

Task Page 60

There are a number of options for these and the candidate may pick from a large range.

Typical answers can include:

1 Petrol stations, fuel and gas depots and processing plants, flour mills and bakeries, wood machine shops, food processing plants, paint spray booths, laboratories, manufacturing premises
2 a. electrostatic discharge, excess voltage
   b. electric shock from discharge
   c. overvoltage damage

SELF ASSESSMENT Page 61

1 b. confirm the fault exists
2 d. the fuse in the 13 A plug has operated
3 c. poor termination
4 d. circuit cable has been damaged
5 b. electrostatic discharge

Progress check Page 62

1 a. Minor Electrical Installation Works Certificate
2 b. Access arrangements
3 a. Courteous
4 d. Manufacturer's instructions
5 b. Take less time
6 c. Maintenance schedule
7 b. Expected delivery date
8 d. 50 mA
9 a. 3
10 d. Locking off the circuit breaker
11 b. Working, using a proving unit
12 b. 4 mm
13 c. Their own unique padlock and key separate from all others
14 a. Electrical Installation Certificate

15 b. Over a period of time
16 c. A fault to earth
17 c. A fault to earth
18 d. Arcing at the switch terminals
19 a. A low insulation resistance fault
20 a. The presence of a second person

Chapter 4

Recap Page 65

- symptoms   exists
- leading   manufacturers   maintenance
- knowledge   circuits   protection   type supply
- restoring   fabric
- supply   reduced   overcurrent   fault malfunction
- 3%   5%
- overload   short   earth
- contacts   arcing   open
- high   transient   failure   short   open
- wiring   terminations   connections accessories   metering
- safety   live   other people   general public   restricted   hazardous
- barriers   open   ladders   access stairwells   live   equipment

Try this Page 72

1 a. low resistance ohmmeter
   b. insulation resistance tester
   c. clamp-on ammeter
   d. voltmeter
2 a. calibration in date
   b. not damaged
   c. battery OK
   d. function

Try this Page 77

1   a. Schedule of Test Results
    b. repair record sheet
    c. MEIWC
2   There may be a number of answers here but these are given in the chapter

Try this Page 82

Expected $R_1 + R_n = \dfrac{m\Omega m \times L}{1000}$

$= \dfrac{14.82 \times 30}{1000} = 0.4446\ \Omega$

the measured value is too high

Try this Page 84

1   seek permission to isolate
2   safely isolate and lock off
3   isolate each fuse spur outlet
4   ensure any other equipment is isolated
5   no connection between live conductors and earth

Try this Page 86

a   Long lead test and the $R_1 + R_2$ test
b   Circuit breaker will not operate to disconnect the supply

Try this Page 89

measured $Zs = 1.15 \times 0.8 = 0.92\ \Omega$. As $0.85\ \Omega < 0.92\ \Omega$ circuit is OK

Try this Page 93

1   300 mA and 300 ms
2   S Type

Try this: Crossword Page 98

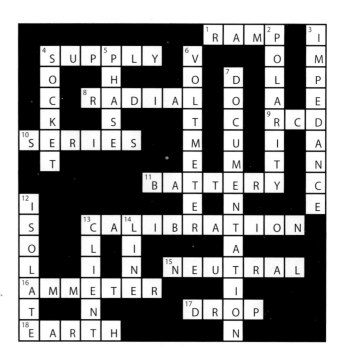

SELF ASSESSMENT Page 99

1   c. insulation resistance ohmmeter
2   d. minor electrical installation works certificate
3   c. 1.51 Ω
4   a. no effect
5   d. ramp test

Chapter 5

Recap Page 100

● line   400 V   line   earth   230 V
● Executive   GS 38   instruments   electricians
● accuracy   national   accurate
● quickly   not   short cuts   safety
● fixed   certification   compliance

- Resistance   nulled   results
- 500   1.0
- single   protective   control   line
- Z_s low   protective   disconnect   time
- additional   fault   risk   fire
- 150
- increasing   trips   current   trip
- rotating disc   indicator
- parallel   series
- Current   clamp-on

Try this Page 107

1   Availability of parts, access, plant and equipment, and labour
2   a. Out of hours working, temporary supplies, restricted access
    b. Age and condition, availability of parts, time for repair vs time for replacement

Try this Page 112

a $\dfrac{0.30 + 0.31}{4} = 0.15\Omega$

b $\dfrac{0.30 + 0.52}{4} = 0.21\Omega$

Try this Page 116

1   measured $Z_s = 1.44 \times 0.8 = 1.152 \ \Omega$
    As $1.2 \ \Omega > 1.152 \ \Omega$ the circuit fails
2   150 mA and 40 ms

Task Page 119

1   MEICW certificate
2   EIC, Schedule of Inspections and Schedule of Test Results

SELF ASSESSMENT Page 128

1   d. the equipment supplier
2   a. low resistance ohmmeter
3   c. 0.24 Ω
4   c. only statement i. is correct
5   b. site instruction

End test

1   b. Manufacturer's instructions
2   a. Minor Electrical Installation Works Certificate
3   c. The type of test instruments used
4   d. Accurate
5   b. Isolation of circuits
6   b. Manufacturer's information
7   a. 50 mA
8   d. 10
9   b. GS 38
10 a. Machines will not function
11 d. Are familiar with the equipment being worked on
12 d. Analyze the information
13 c. The circuit protective device has operated
14 a. Upstream
15 d. Arcing
16 c. Faulty starter
17 b. An overload of the circuit
18 b. Low resistance ohmmeter
19 b. Disconnect the circuit at the mid point and test each half
20 a. Check the voltage with an approved voltage indicator
21 b. Clamp meter
22 b. Short circuit
23 a. Minor Electrical Installation Works Certificate
24 a. Time taken to rectify a fault
25 c. Continuity of cpc
26 b. 0.25 Ω
27 a. Both statements are correct
28 b. Earth fault loop impedance
29 d. Delay
30 c. Fire rated

Glossary

AVI approved voltage indicator

cpc circuit protective conductor

DNO distribution network operator

ECA Electrical Contractors Association

Efficacy the ability to produce a desired amount of a desired effect

EIC Electrical Installation Certificate

EICR Electrical Installation Condition Report

emf electromotive force

ES Edison screw

ESD electrostatic discharge

ESQCR Electricity Safety, Quality and Continuity Regulations

EWR Electricity at Work Regulations

FELV functional extra-low voltage

HSE Health and Safety Executive

IET Institution of Engineering and Technology

IT information technology

L line

LED light emitting diode

LV low voltage

MEIWC Minor Electrical Installation Works Certificate

MET main earthing terminal

PELV protective extra-low voltage

PIR Periodic Inspection Report

PLC programmable logic controller

PME protective multiple earthing. See TN-C-S

PPE personal protective equipment

RCD residual current device

SELV separated extra-low voltage

Stroboscopic effect intense flashing or pulsing light of various frequencies affecting visual effects of rotary motion

TN-C-S A supply system in which the earth provision is provided by the DNO using a combined neutral and earth conductor within the supplier's network cables. The earth and neutral are then separated throughout the installation. These systems are referred to as TN-C-S or PME systems

TN-S A supply system in which the earth provision is provided by the DNO using a separate metallic conductor provided by the Distribution Network Operator (DNO). This provision may be by connection to the metal sheath of the supply cable or a separate conductor within the supply cable

TRS tough rubber sheath

TT a supply system in which the DNO does not provide an earth facility. The installation's exposed and extraneous metalwork is connected to earth by a separate installation earth electrode and uses the general mass of earth as the return path

VRI vulcanized rubber insulated

Z_e earth fault loop impedance

Z_s system earth fault loop impedance

Index: EIS Fault Finding and Diagnosis (Doughton and Hooper)